A Guide to Poisonous House and Garden Plants

A Guide to Poisonous House and Garden Plants

Anthony P. Knight
BVSc., MS, DACVIM
Professor
Department of Clinical Sciences
College of Veterinary Medicine and Biomedical Sciences
Colorado State University

TETON NEWMEDIA
INNOVATIVE PUBLISHING OF VETERINARY & HUMAN MEDICINE

Executive Editor: Carroll C. Cann
Development Editor: Susan L. Hunsberger

Teton NewMedia
P.O. Box 4833
Jackson, WY 83001

1-888-770-3165
tetonnewmedia.com

Printed in the United States of America

Print Number 5 4 3 2 1

Library of Congress Cataloging-in-Publication Data on File

ISBN # 1-59161-028-1

Dedication and Acknowledgments

To my wife Cassandra, thanks for your encouragement and support during the preparation of this book, and for the endless hours you waited while I photographed the plants included in the book.

Special thanks go to former students who gave impetus to the book's inception.

Special recognition and thanks go to the following who provided photographs of plants for this book:

Catherine Trendell, Painswick Rd, Gloucester England

John and Emily Smith, Baldwin, Georgia

Art Whistler, 2814 Kalawao St. Honolulu HI, 96822

Julie Loquidis and Brinsley Burbidge, 127 Estate St. George, Frederiksted Virgin Islands 00840

E. Vinter, University of Pretoria, Onderstepoort, South Africa

Virginia L. Smith, College of Veterinary Medicine and Biomedical Sciences Colorado State University, Fort Collins 80523

Larry Allain, National Wetlands Research Center 700 Cajundome Bldv, Lafayette, LA 70506

Glen Lee, Saskatchewan, Canada

Disclaimer

The poisonous plants covered in this book are those that are commonly and not so commonly grown as house and/or garden plants in North America and other areas of the world. The list of poisonous plants is not exhaustive, but rather the book attempts to provide information on those plants that have been documented as toxic to animals and people, or have good potential for poisoning. Undoubtedly, more plants will be added to those in this book as new plant species are introduced to the environments in which house hold plants live! The author has made every effort to use botanically accurate names for plants although the ever changing taxonomy of plant names can make it difficult to satisfy all taxonomists!

Although the book gives general treatment guidelines for each of the plants, specific details of appropriate drugs, their doses, and supportive therapy are not given because of the great variability in the patient's history, and symptoms. Where plant poisoning is suspected it is critical that a Poison Control Center such as the National Animal Poison Control Center (1 888 4ANI HELP), and/or a knowledgeable veterinarian be consulted as soon as possible.

Contents

While there are well illustrated books covering the subject of plant poisoning in people, such books are lacking on current information on plant toxicity in dogs and cats that can serve as a resource for veterinarians, pet owners, landscapers, plant nurseries, and others.

Plant poisoning of animals and people continues to be a frequently encountered problem in today's home and garden environment. So much so that poison control centers receive thousands of calls every year regarding plant poisoning. As stated in the American Association of Poison Control Centers 2003 annual report, over 57,000 questions regarding plant poisoning of children were received by poison control centers in 2003.[1] The most frequent plant exposures from this study are shown in Table 1. In another study from 1994-2001 involving 260,235 cases of poisoning exposures in children from 6-12 years of age, it was determined that 4% of the cases were attributed to plant poisoning.[2] The majority of cases were attributed to household products (20%), office products (14%), foreign bodies (12%), over-the-counter medicines (10%), cosmetics (9%), prescription medications (8%), and industrial chemicals (6%).[2] Based upon the 10 most frequently reported signs of poisoning in people associated with 776,284 plant exposure cases reported to poison control centers in the United States, 20 common plants were most commonly ingested. (Table 2)[3] Similarly in a study of poison exposures in dogs and cats reported to poison control centers from 1993-1994, of the 116,432 poison exposures reported in dogs, 10.97% were attributed to plants, and of the 19,489 poison exposures in cats, 25.24% were plant related.[4]

Owing to the international trade of plant species for commercial purposes, more and more plants are being introduced as ornamentals to the home and garden that are toxic to people and animals. For example the blood lily (*Haemanthus* spp.), glory lily *(Gloriosa superba)*, and kalanchoes (*Kalanchoe* spp.) are well known toxic plants in their native habitat in Africa and are now commonly available as garden and house plants in many parts of the world.

This book is a resource for helping identify poisonous plants in the home and garden, as well as providing information on the toxic properties of the plants and the clinical signs that can be expected in animals that eat the plants. The book is organized alphabetically by the plant's botanical name. In the index, common names are also listed to help in finding the specific plant in question. For each plant, the common names, most common species and a description of the plant along with one or more pictures of the plant are provided. Emphasis has been placed on providing plant pictures that help the reader quickly identify the plant. A glossary of terms is also provided for clarity. Wherever possible, the principle toxin(s) present in the plant, its mode of action and the clinical signs of poisoning it will cause in animals are covered. For each plant the author has given a brief assessment of the relative risk that the plant poses to household pets, and where relevant the potential risk to people. While most plant poisoning exposures in dogs and cats cause relatively mild and transient signs, general treatment recommendations are given for each plant. It is beyond the scope of this book to provide the spectrum of treatment possibilities that are available, and it is therefore important that a veterinarian be consulted when large quantities of plant material have been or are suspected of having been eaten, and if the animal appears in distress. Early treatment can make all the difference in the animal's recovery. This is especially true in cats that have chewed and eaten the leaves or flowers of lilies such as the Easter lily.

The famous statement by Paracelsus, born Theophrastus Phillippus Aureolus Bombastus von Hohenheim in Switzerland, aptly summarizes the basic concept of what makes a plant poisonous.
"Alle Ding' sind Gift und nichts ohn' Gift; allein die Dosis macht, das ein Ding kein Gift ist", which translated states "All things are poison and nothing (is) without poison; only the dose makes that a thing is no poison." In other words, "the dose makes the poison"

Paracelsus (1493-1541)

As the reader will discover in this book there are many beautiful flowering plants available for the house and garden that are potentially harmful. It is the author's sincere hope that the reader will not be deterred from enjoying these plants in their home and garden provided common sense prevails!

Preface Table 1
Frequency of Reported Human Plant Exposures

Genus	Common Name	Exposures
Spathiphyllum	Peace lily	3,602
Philodendron	Philodendron	2,880
Euphorbia pulcherrima *	Poinsettia	2,620
Ilex	Holly	2,427
Phytolacca	Pokeweed	1,863
Ficus	Fig	1,612
Toxicodendron	Poison ivy/oak	1,500
Dieffenbachia	Dumb cane	1,324
Crassula	Jade plant	1,146
Epipremnum	Pothos	1,083
Capsicum annuum	Chili pepper	1,049
Rhododendron	Azalea, rhododendron	1,047
Chrysanthemum	Chrysanthemum	869
Nerium oleander	Oleander	847
Schlumbergera bridgessi	Christmas cactus	841
Hedera helix	Ivy	769
Eucalyptus	Eucalyptus	727
Malus *	Apple	703
Nandina domestica	Heavenly bamboo	694
Saintpaulia ionantha *	African violet	685

* Minimally or not toxic.

Table compiled from: Watson WA et al: 2002 annual report of the American Association of Poison Control Centers Toxic Exposure Surveillance System. Am J Emergency Medicine 2003, 21:353-421.

Preface Table 2
The 20 Plants Most Commonly Ingested by People

Genus	Common Name
1. *Capsicum*	Pepper
2. *Dieffenbachia*	Dumb cane
3. *Philodendron*	Philodendron
4. *Caladium*	Caladium
5. *Spathiphyllum*	Peace lily
6. *Datura*	Jimson weed
7. *Euphorbia*	Spurge
8. *Ficus*	Fig
9. *Narcissus*	Daffodil
10. *Phytolacca*	Ink berry
11. *Zantedeschia*	Calla lily
12. *Ilex*	Holly
13. *Iris*	Iris
14. *Epipremnum*	Pothos
15. *Eucalyptus*	Eucalyptus
16. *Ephedra*	Mormon Tea
17. *Hedera*	Ivy
18. *Arisaema*	Jack in the pulpit
19. *Piper nigrum*	Black pepper
20. *Begonia*	Begonia

Table compiled from: Mrvos R. Krenzelok EP, Jacobsen TD: Toxidromes associated with the most common plant ingestions. Vet Human Toxicol 2001, 43:366-369.

References

1. Watson WA et al: 2002 annual report of the American Association of Poison Control Centers Toxic Exposure Surveillance System. Am J Emergency Medicine 2003, 21: 353-421.

2. Vassilev ZP. Marcus S. Jennis T. Ruck B. Rego G. Swenson R. Halperin W: Trends in major types of poisoning exposures in children reported to a regional poison control center, 1994-2001. Clin Pediatrics 2004, 43: 573-576.

3. Mrvos R, Krenzelok EP, Jacobsen TD: Toxidromes associated with the most common plant ingestions. Vet Human Toxicol 2001, 43: 366-369.

4. Hornfeldt CS, Murphy MJ: American Association of Poison Control Centers report on poisonings of animals, 1993-1994. J Am Vet Med Assoc 1998, 212: 358-361.

Table I
Plants Grouped According to their Primary Toxins

Alkaloids

Genus	Common Name	General Effect
Aconitum spp.	Monk's hood, aconite	Vomiting, death
Albizia spp.	Mimosa tree	Neurologic signs
Aleurites spp.	Tung nut	Gastroenteritis
Amaryllis belladonna	Amaryllis, naked lady	Vomiting, diarrhea
Atropa belladonna	Deadly nightshade	Tachycardia,mydriasis
Baptisia spp.	Indigo	Muscle tremors, vomiting
Brugmansia spp.	Angel's trumpet	Tachycardia, respiratory failure
Brunfelsia spp.	Yesterday-today-tomorrow	Seizures, vomiting
Buxus spp.	Boxwood	Vomiting, diarrhea
Calycanthus spp.	Sweet shrub, allspice	Muscle tremors, seizures
Cassia spp.	Golden shower tree	Gastroenteritis
Catharanthus spp.	Periwinkle	Neurotoxic, hypotension
Celastrus spp.	American bittersweet	Vomiting, diarrhea
Cestrum spp.	Cestrum, jasmine	Hepatotoxicity, calcinosis
Chelidonium majus	Celandine	Vomiting, diarrhea, seizures
Clivia spp.	Clivia	Vomiting, diarrhea, seizures
Colchicum spp.	Autumn crocus	Gastroenteritis, shock
Conium maculataum	European, spotted hemlock	Neurotoxic, teratogenic
Corydalis spp.	Fitweed, corydalis	Vomiting, muscle tremors
Crinum spp.	Crinum lily	Vomiting, diarrhea, hypotensive
Datura spp.	Sacred datura, Jimson weed	Respiratory failure, hallucinations
Dicentra spp.	Bleeding heart	Vomiting, muscle tremors
Erythrina spp.	Coral bean	Muscle paralysis
Euonymus spp.	Burning bush	Gastroenteritis, cardiac dysrhythmias

continued

Table I Continued

Genus	Common Name	General Effect
Galanthus spp.	Snow drop	Vomiting, diarrhea
Gelsemium spp.	Carolina jessamine	Neurotoxic, death
Gloriosa superba	Glory lily	Gastroenteritis, shock
Haemanthus spp.	Blood lily	Vomiting, diarrhea, seizures
Heliotropium spp.	Heliotrope	Liver failure
Hippeastrum spp.	Amaryllis lily	Vomiting, diarrhea,
Hymenocallis spp.	Spider lily, Peruvian daffodil	Gastroenteritis
Hyoscyamus spp.	Black henbane	Tachycardia, respiratory failure
Ilex spp.	Holly	Gastroenteritis
Ipomoea spp.	Morning glory	Neurologic abnormalities
Laburnum spp.	Golden chain tree	Vomiting, tachycardia
Ligularia spp.	Ligularia	Liver failure
Lobelia spp.	Cardinal flower	Vomiting, cardiotoxicity
Mirabilis spp.	Four o'clocks	Gastroenteritis
Nandina domestica	Heavenly bamboo	Seizures, respiratory failure
Narcissus spp.	Daffodils, narcissus	Vomiting, diarrhea
Nerine spp.	Nerine, or spider lily	Vomiting, diarrhea
Nicotiana spp.	Tobacco	Seizures, respiratory failure
Papaver spp.	Poppy	Excitement, depression
Physalis spp.	Ground cherry	Vomiting, colic
Sanguinaria canadensis	Blood root	Vomiting, colic
Senecio (Packera) spp.	Senecio, groundsel	Liver failure
Solandra spp.	Chalice vine	Respiratory failure, hallucinations
Solanum pseodocapsicum	Jerusalem cherry	Vomiting, respiratory failure
Sophora spp.	Mescal bean	Neurologic signs, ataxia
Sprekelia spp.	Maltese cross lily	Vomiting, diarrhea, ataxia
Symphytum spp.	Comfrey	Liver failure
Taxus spp.	Yew	Cardiac dysrhythmias, death

Genus	Common Name	General Effect
Theobroma cocoa	Cocoa	Vomiting, diarrhea, neurologic signs
Zephyranthes spp.	Atamasco or rain lily	Vomiting, gastroenteritis

Cannabinoids

Cannabis sativa	Marijuana, hemp	Neurologic signs

Amino Acids

Mimosa spp.	Sensitive plant	Hair loss, hyperthyroidism
Zamia spp.	Sago palm	Neurologic signs

Toxalbumins (Lectins), Proteins

Abrus precatorius	Rosary pea	Enteritis, shock, death
Blighia sapida	Akee	Hypoglycemia, seizures
Hyacinthus spp.	Hyacinth	Gastroenteritis, death
Jatropha spp.	Physic nut, coral plant	Gastrointestinal irritation
Phoradendron spp.	Mistletoe	Gastrointestinal irritation
Ricinus communis	Castor bean	Gastroenteritis, shock, death
Robinia spp.	Black locust	Gastroenteritis, shock
Tulipa spp.	Tulip	Gastroenteritis, death
Wisteria spp.	Wisteria	Gastrointestinal irritation

Oxalates

Aglaonema spp.	Chinese evergreen	Salivation, oral edema
Alocasia spp.	Elephant's ears	Salivation, oral edema
Anthurium spp.	Flamingo flower	Salivation, oral edema
Arum spp.	Arum	Salivation, oral edema
Begonia spp.	Begonia	Salivation, oral edema
Caladium spp.	Caladium	Salivation, oral edema
Calla palustris	Water arum	Salivation, oral edema
Caryota spp.	Fish tail palm	Salivation, oral edema
Colocasia spp.	Elephant's ears, taro	Salivation, oral edema
Dieffenbachia spp.	Dumb cane	Salivation, oral edema
Epipremnum spp.	Pothos	Salivation, oral edema

continued

Table I Continued

Genus	Common Name	General Effect
Monstera spp.	Ceriman, Swiss cheese plant	Salivation, oral edema
Narcissus spp.	Daffodils, narcissus	Salivation, oral edema
Oxalis spp.	Oxalis, sorrel	Salivation, hypocalcemia
Parthenocissus spp.	Virginia creeper	Vomiting, diarrhea
Philodendron spp.	Philodendron	Salivation, oral edema
Phytolacca american	Pokeberry	Gastroenteritis
Pistia stratiotes	Water lettuce	Salivation, oral edema
Portulaca spp.	Moss rose	Salivation, vomiting
Spathiphyllum spp	Peace lily	Salivation, oral edema
Syngonium spp.	Arrow head vine	Salivation, oral edema
Zantedeschia spp.	Calla, arum lily	Salivation, oral edema

Glycosides

Genus	Common Name	General Effect
Acocanthera spp.	Bushman's poison	Dysrhythmias, death
Actea spp.	Bane berry	Gastroenteritis
Adenium obesum	Desert rose	Cardiac dysrhythmias
Adonis spp.	Adonis, pheasant's eye	Cardiac dysrhythmias, enteritis
Aesculus spp.	Buckeye, horse chestnut	Vomiting, diarrhea, neurologic signs
Aloe spp.	Aloe	Diarrhea
Anemone spp.	Anemone	Gastroenteritis
Asclepias spp.	Milkweed	Cardiac dysrhythmias, neurotoxicity
Aucuba spp.	Aucuba, gold dust tree	Vomiting, gastroenteritis
Bowiea spp.	Climbing onion	Cardiac dysrhythmias
Calotropis spp.	Crown flower	Cardiac dysrhythmias, neurotoxicity
Caltha spp.	Marsh marigold	Gastroenteritis
Clematis spp.	Clematis	Gastroenteritis, diarrhea
Convallaria majalis	Lily of the valley	Cardiac dysrhythmias, enteritis
Coriaria spp.	Coriaria	Gastroenteritis, neurotoxic
Crassula spp.	Jade plant	Cardiac dysrhythmias, gastroenteritis
Cryptostegia spp.	Rubber vine	Gastroenteritis

Genus	Common Name	General Effect
Cycas spp.	Cycad, sago palm	Liver necrosis, neurotoxicity
Digitalis spp.	Fox glove	Cardiac dysrhythmias, gastroenteritis
Eriobotrya spp.	Loquat	Cyanosis, respiratory failure, death
Euonymus spp.	Burning bush	Gastroenteritis, cardiac dysrhythmias
Euonymus spp.	Euonymus	Vomiting, gastroenteritis, dysrhythmias
Helleborus spp.	Christmas rose, hellebore	Gastroenteritis, cardiac dysrhythmias
Hoya spp	Wax plant	Cardiac dysrhythmias
Hyacinthoides spp.	English bluebell	Cardiac dysrhythmias, gastroenteritis
Hyacinthus spp.	Hyacinth	Vomiting, diarrhea
Hydrangea spp.	Hydrangea	Cyanosis, respiratory failure, death
Kalanchoe spp.	Kalanchoe	Cardiac dysrhythmias, gastroenteritis
Nandina domestica	Heavenly bamboo	Cyanosis, respiratory failure, death
Nerium oleander	Oleander	Cardiac dysrhythmias, death
Ornithogallum spp.	Star of Bethlehem	Cardiac dysrhythmias
Podophylum spp.	May apple	Gastroenteritis, neurologic signs
Prunus spp.	Choke cherry	Cyanosis, respiratory failure, death
Pyracantha coccinea	Pyracantha	Cyanosis, respiratory failure
Ranunculus spp.	Buttercups	Gastroenteritis
Scilla spp.	Scilla, sqill	Gastroenteritis, cardiac dysrhythmias
Thalictrum spp.	Meadow rue	Gastroenteritis
Thevetia spp.	Yellow oleander	Cardiac dysrhythmias, death
Zamia spp.	Sago palm	Liver necrosis

continued

Table 1 Continued

Genus Saponins/Sapogenins	Common Name	General Effect
Aesculus spp.	Buckeye, horse chestnut	Gastroenteritis
Agapanthus spp.	Blue lily	Gastroenteritis
Agave lecheguilla	Agave	Cholecystitis, biliary obstruction
Asparagus spp.	Asparagus	Gastroenteritis
Cyclamen spp.	Cyclamen	Gastroenteritis, Cardiac dysrhythmias
Dracaena spp.	Dracaena, corn plant	Gastroenteritis
Gymnocladus spp.	Kentucky coffee tree	Gastroenteritis
Hosta spp.	Hosta	Gastroenteritis
Pedilanthus spp.	Slipper flower	Gastroenteritis
Phytolacca americana	Pokeberry	Gastroenteritis
Pittosporum spp.	Pittosporum	Gastroenteritis
Schefflera spp.	Umbrella tree	Gastroenteritis
Sesbania spp.	Rattlebush, coffee bean	Gastroenteritis
Yucca spp.	Yucca	Cholecystitis, Biliary obstruction

Terpenoids

Acalypha spp.	Chenille plant, copper leaf	Gastroenteritis, dermatitis
Achillea spp.	Yarrow	Gastroenteritis
Allamanda spp.	Allamanda	Gastrointestinal irritation
Caesalpinia spp.	Peacock flower	Gastrointestinal irritation
Capsicum spp.	Peppers	Gastrointestinal irritation
Daphne spp.	Daphne	Gastrointestinal irritation
Delphinium spp.	Larkspur	Neuromuscular paralysis
Euphorbia spp.	Euphorbias	Stomatitis, vomiting
Hedera spp.	Ivy	Gastrointestinal irritation
Iris spp.	Iris	Gastrointestinal irritation

Genus	Common Name	General Effect
Jatropha spp.	Physic nut, coral plant	Gastrointestinal irritation
Kalmia spp.	Laurel	Vomiting, cardiac dysrhythmias
Lantana spp.	Lantana	Cholecystitis/stasis
Leucothoe spp.	Fetterbush	Vomiting, cardiac dysrhythmias
Ligustrum spp.	Privet	Vomiting, diarrhea, ataxia
Lonicera spp.	Honeysuckle	Vomiting, diarrhea
Lyonia spp.	Staggerbush, maleberry	Vomiting, cardiac dysrhythmias
Melaleuca spp.	Melaleuca, tea tree	Neurologic signs
Melia azadarach	Chinaberry	Vomiting, diarrhea, ataxia
Momordica spp.	Balsam pear	Vomiting, diarrhea
Nepeta cataria	Catnip	Abnormal behavior, hallucinations
Pieris spp.	Pieris	Vomiting, cardiac dysrhythmias
Plumeria spp.	Plumeria	Vomiting, diarrhea
Sambucus spp.	Elderberry	Gastrointestinal irritation
Sapindus spp.	Soapberry	Gastrointestinal irritation
Schefflera spp.	Umbrella tree	Salivation, gastroenteritis
Schinus spp.	Pepper tree	Dermatitis, vomiting

Sulfoxides and other Oxidants

Acer rubrum	Red maple	Heinz body anemia, death
Allium species	Onions	Heinz body anemia

Glucosinolates

Armoracia rusticana	Horse raddish	Gastrointestinal irritant

Naphthaquinone

Juglans nigra	Black walnut	Laminits

continued

Table I Continued

Genus	Common Name	General Effect
Acetylenic alcohols (unsaturated)		
Cicuta spp.	Water hemlock	Paralysis, cardiomyopathy
Ketone Benzofurans		
Eupatorium spp.	White snake root	muscle tremors, cardiomyopathy
Phenolics		
Humulus lupulus	Hops	Malignant hyperthermia, vomiting

Table 2
Plants Listed by their Primary Clinical Effects

Plants Causing Excessive Salivation

Genus	Common Name	Toxin
Actea spp.	Bane berry	Ranunculin
Aglaonema spp.	Chinese evergreen	Calcium oxalate raphides
Alocasia spp.	Elephant's ears	Calcium oxalate raphides
Anthurium spp.	Flamingo flower	Oxalates
Arisaema spp.	Jack in the pulpit	Calcium oxalate raphides
Arum spp.	Arum	Calcium oxalate raphides
Begonia spp.	Begonia	Oxalates
Caladium spp.	Caladium	Oxalates
Calla palustris	Water arum	Oxalates
Caltha spp.	Marsh marigold	Ranunculin
Capsicum spp.	Peppers	Capsaicin
Caryota spp.	Fish tail palm	Calcium oxalate raphides
Colocasia spp.	Elephant's ears, taro	Calcium oxalate raphides
Daphne spp.	Daphne	Diterpenes
Dieffenbachia spp.	Dumb cane	Calcium oxalate raphides
Dracaena spp.	Dracaena, corn plant	Saponins
Epipremnum spp.	Pothos	Calcium oxalate raphides
Euphorbia spp.	Euphorbias	Diterpenoids (euphorbol esters)
Monstera spp.	Ceriman, Swiss cheese plant	Calcium oxalate raphides
Narcissus spp.	Daffodils, narcissus	Phenanthridine alkaloids, oxalates
Oxalis spp.	Oxalis, sorrel	Oxalates
Philodendron spp.	Philodendron	Calcium oxalate raphides
Pistia stratiotes	Water lettuce	Calcium oxalate raphides

continued

Table 2 Continued

Genus	Common Name	Toxin
Ranunculus spp.	Buttercups	Ranunculin
Spathiphyllum spp.	Peace lily	Calcium oxalate raphides
Sprekelia spp.	Maltese cross lily	Phenanthadine alkaloids
Syngonium spp.	Arrow head vine	Calcium oxalate raphides
Zantedeschia spp.	Calla, arum lily	Calcium oxalate raphides

Plants Causing Vomiting

Abrus precatorius	Rosary pea	Lectins (abrin)
Aconitum spp.	Monk's hood, aconite	Diterpenoid alkaloids
Actea spp.	Bane berry	Ranunculin
Aesculus spp.	Buckeye, horsechestnut	Esculin, escin
Aleurites spp.	Tung nut	Diterpenoids
Amaryllis belladonna	Amaryllis, naked lady	Phenanthridine alkaloids
Anemone spp.	Anemone	Ranunculin
Araceae family	(Various)	Oxalates
Baptisia spp.	Indigo	Quinolizidine alkaloids (cytisine)
Blighia sapida	Akee	Amino acid Hypoglycin A,B
Brunfelsia spp.	Yesterday-today-tomorrow	Brunfelsamidine
Buxus spp.	Boxwood	Steroidal alkaloids
Caesalpinia spp.	Peacock flower	Diterpenoids, gallotannins
Calotropis spp.	Crown flower	Cardenolides
Caltha spp.	Marsh marigold	Ranunculin
Chelidonium majus	Celandine	Berberine, chelidonine,
Clivia spp.	Clivia	Phenanthridine alkaloids (lycorine)
Colchicum spp.	Autumn crocus	Colchicine
Convallaria majalis	Lily of the valley	Cardenolides
Coriaria spp.	Coriaria	Coriamyrtin
Corydalis spp.	Fitweed, corydalis	Apomorphine, protoberberine
Crassula spp.	Jade plant	Cardenolides

Genus	Common Name	Toxin
Crinum spp.	Crinum lily	Phenanthridine alkaloids (lycorine)
Cycas spp.	Cycad, sago palm	Glycosides
Cyclamen spp.	Cyclamen	Terpenoid saponins
Daphne spp.	Daphne	Diterpenes
Dicentra spp.	Bleeding heart	Isoquinoline alkaloids
Digitalis spp.	Fox glove	Digitalis glycosides
Dracaena spp.	Dracaena, corn plant	Saponins
Eucharis spp.	Eucharist lily	Phenanthridine alkaloids
Euphorbia spp.	Euphorbias	Diterpenoids (euphorbol esters)
Galanthus spp.	Snow drop	Phenanthridine alkaloids
Gloriosa superba	Glory lily	Colchicine
Haemanthus spp.	Blood lily	Phenanthridine alkaloids
Hedera spp.	Ivy	Triterpenoid saponins
Hippeastrum spp.	Amaryllis lily	Phenanthridine alkaloids
Hosta spp.	Hosta, plantain lily	Saponins
Humulus lupulus	Hops	Phenolic compounds
Hyacinthoides spp.	English bluebell	Cardenolides
Hyacinthus spp.	Hyacinth	Tuliposide, lectin
Kalmia spp.	Laurel	Diterpenoids (Grayanotoxins)
Lonicera spp.	Honeysuckle	Triterpenoid saponins
Lyonia spp.	Staggerbush, maleberry	Diterpenoids (Grayanotoxins)
Macadamia spp.	Macadamia nut	Unknown toxin
Melia azadarach	Chinaberry	Triterpines (meliatoxins)
Momordica spp.	Balsam pear	Cucurbitane terpenoids
Narcissus spp.	Daffodils, narcissus	Phenanthridine alkaloids, oxalates
Nerine spp.	Nerine, or spider lily.	Phenanthridine alkaloids
Parthenocissus spp.	Virginia creeper	Oxalates, other
Pedilanthus spp.	Slipper flower	Saponins
Philodendron spp.	Philodendron	Calcium oxalate raphides

continued

Table 2 Continued

Genus	Common Name	Toxin
Phoradendron spp.	Mistletoe	Lectins
Physalis spp.	Ground cherry	Glycoalkaloids (solanine)
Phytolacca american	Pokeberry	Saponins, Oxalates
Pieris spp.	Pieris	Diterpenoids (Grayanotoxins)
Pistia stratiotes	Water lettuce	Calcium oxalate raphides
Pittosporum spp.	Pittosporum	Saponins
Plumeria spp.	Plumeria	Terpenoids
Podocarpus spp.	Buddhist or yew pine	Unknown
Ranunculus spp.	Buttercups	Ranunculin
Rhododendron spp.	Rhododendron, azalea	Diterpenoids (Grayanotoxins)
Ricinus communis	Castor bean	Ricin
Robinia spp.	Black locust	Lectins (robinin)
Sambucus spp.	Elderberry	Triterpenoids
Sanguinaria canadensis	Blood root	Sanguinarine alkaloids
Schefflera spp.	Umbrella tree	saponins, terpenoids, oxalates
Schinus spp.	Pepper tree	Terpenoids
Scilla spp.	Scilla, sqill	Bufadienolides
Solenostemon spp.	Coleus	Irritant oil Coleon O.
Sprekelia spp.	Maltese cross lily	Phenanthadine alkaloids
Thalictrum spp.	Meadow rue	Ranunculin
Theobroma cocoa	Cocoa	Methylxanthine alkaloids
Tulipa spp.	Tulip.	Tuliposide, and lectins
Wisteria spp.	Wisteria	Glycoproteins (lectins)
Yucca spp.	Yucca	Saponins
Zephyranthes spp.	Atamasco or rain lily	Phenanthridine alkaloids

Plants Causing Diarrhea

Genus	Common Name	Toxin
Abrus precatorius	Rosary pea	Lectins (abrin)
Achillea spp.	Yarrow	Mono-, sesquiterpenes
Aconitum spp.	Monk's hood, aconite	Diterpenoid alkaloids
Actea spp.	Bane berry	Ranunculin
Adonis spp.	Adonis, pheasant's eye	Ranunculin

Genus	Common Name	Toxin
Aesculus spp.	Buckeye, horsechestnut	Esculin, escin
Aleurites spp.	Tung nut	Diterpenoids
Allamanda spp.	Allamanda	Terpenoids, iridoids
Aloe spp.	Aloe	Anthraquinone glycosides
Anemone spp.	Anemone	Ranunculin
Armoracia rusticana	Horse raddish	Glucosinolates (sinigrin)
Asparagus spp.	Asparagus	Sapogenins
Aucuba spp.	Aucuba, gold dust tree	Iridoid glycoside (Aucubin)
Buxus spp.	Boxwood	Steroidal alkaloids
Caesalpinia spp.	Peacock flower	Diterpenoids, gallotannins
Calotropis spp.	Crown flower	Cardenolides
Caltha spp.	Marsh marigold	Ranunculin
Capsicum spp.	Peppers	Capsaicin
Cassia spp.	Golden shower tree	Anthraquinones
Celastrus spp.	American bittersweet	Sesquiterpene alkaloids (celastrine)
Chelidonium majus	Celandine	Berberine, chelidonine
Clematis spp.	Clematis	Ranunculin
Clivia spp.	Clivia	Phenanthridine alkaloids (lycorine)
Colchicum spp.	Autumn crocus	Colchicine
Convallaria majalis	Lily of the valley	Cardenolides
Crinum spp.	Crinum lily	Phenanthridine alkaloids (lycorine)
Cryptostegia spp.	Rubber vine	Cardenolides
Cyclamen spp.	Cyclamen	Terpenoid saponins
Daphne spp.	Daphne	Diterpenes
Digitalis spp.	Fox glove	Digitalis glycosides
Eucharis spp.	Eucharist lily	Phenanthridine alkaloids
Euonymus spp.	Burning bush	Cardenolides, alkaloids
Euonymus spp.	Euonymus	Cardenolides, saponins
Galanthus spp.	Snow drop	Phenanthridine alkaloids
Gloriosa superba	Glory lily	Colchicine

continued

Table 2 Continued

Genus	Common Name	Toxin
Gymnocladus spp.	Kentucky coffee tree	Gymnocladosapponins
Haemanthus spp.	Blood lily	Phenanthridine alkaloids
Hedera spp.	Ivy	Triterpenoid saponins
Helleborus spp.	Christmas rose, hellebore	Cardenolides, ranunculin
Hibiscus spp.	Hibiscus	Unknown toxin
Hippeastrum spp.	Amaryllis lily	Phenanthridine alkaloids
Hosta spp.	Hosta, plantain lily	Saponins
Hyacinthoides spp.	English bluebell	Cardenolides
Hyacinthus spp.	Hyacinth	Tuliposide, and lectins
Kalanchoe spp.	Kalanchoe	Cardenolides
Lobelia spp.	Cardinal flower	Pyridine alkaloids (lobeline)
Lonicera spp.	Honeysuckle	Triterpenoid saponins
Melia azadarach	Chinaberry	Triterpines (meliatoxins)
Mirabilis spp.	Four o'clocks	Alkaloids (trigonelline)
Momordica spp.	Balsam pear	Cucurbitane terpenoids
Narcissus spp.	Daffodils, narcissus	Phenanthridine alkaloids, oxalates
Nerine spp.	Nerine, or spider lily	Phenanthridine alkaloids
Parthenocissus spp.	Virginia creeper	Oxalates, other
Pedilanthus spp.	Slipper flower	Saponins
Phoradendron spp.	Mistletoe	Lectins
Phytolacca americana	Pokeberry	Saponins, oxalates
Pieris spp.	Pieris	Diterpenoids (Grayanotoxins)
Pittosporum spp.	Pittosporum	Saponins
Plumbago spp.	Plumbago	Naphthaquinones
Podocarpus spp.	Buddhist or yew pine	Unknown
Podophylum spp.	May apple	Podophyllotoxin
Ranunculus spp.	Buttercups	Ranunculin
Rhamnus spp.	Buckthorn	Anthraquinones
Rhododendron spp.	Rhododendron, azalea	Diterpenoids (Grayanotoxins)
Ricinus communis	Castor bean	Ricin

Genus	Common Name	Toxin
Robinia spp.	Black locust	Lectins (robinin)
Sambucus spp.	Elderberry	Triterpenoids
Sapindus spp.	Soapberry	Triterpenoids
Schefflera spp.	Umbrella tree	Saponins, terpenoids, oxalates
Scilla spp.	Scilla, sqill	Bufadienolides
Sesbania spp.	Rattlebush, coffee bean	Saponins
Sprekelia spp.	Maltese cross lily	Phenanthadine alkaloids
Thalictrum spp.	Meadow rue	Ranunculin
Theobroma cocoa	Cocoa	Methylxanthine alkaloids
Tulipa spp.	Tulip	Tuliposide, and lectins
Wisteria spp.	Wisteria	Glycoproteins (lectins)
Yucca spp.	Yucca	Saponins
Zephyranthes spp.	Atamasco or rain lily	Phenanthridine alkaloids

Plants Causing Renal Failure

Lilium species	Lilies	Unknown toxin
Sesbania spp.	Rattlebush, coffee bean	Saponins
Vitis spp.	Grapes	Unknown toxin

(All oxalate containing plants have the potential to cause renal failure)

Plants Causing Hemolysis, Methemoglobinemia, and Anemia

Acer rubrum	Red maple	Oxidants, gallic acid
Allium species	Onions	Alkylcysteine sulfoxides
Catharanthus spp.	Periwinkle	Vincristine, reserpine

Plants Causing Cardiac Abnormalities

Acocanthera spp.	Bushman's poison	Cardiac glycosides
Aconitum spp.	Monk's hood, aconite	Diterpenoid alkaloids
Adenium obesum	Desert rose	Cardiac glycosides
Adonis spp.	Adonis, pheasant's eye	Strophanthidin,
Asclepias spp.	Milkweed (broadleaf)	Cardenolides
Atropa belladonna	Deadly nightshade	Tropane alkaloids (hyoscyamine)
Bowiea spp.	Climbing onion	Bufadienolides

continued

Table 2 Continued

Genus	Common Name	Toxin
Brugmansia spp.	Angel's trumpet	Tropane alkaloids (hyoscyamine)
Calotropis spp.	Crown flower	Cardenolides
Cicuta spp.	Water hemlock	Cicutoxin, cicutol
Colchicum spp.	Autumn crocus	Colchicine
Convallaria majalis	Lily of the valley	Cardenolides
Crassula spp.	Jade plant	Cardenolides
Cryptostegia spp.	Rubber vine	Cardenolides
Cyclamen spp.	Cyclamen	Terpenoid saponins
Datura spp.	Sacred datura, Jimson weed	Tropane alkaloids
Digitalis spp.	Fox glove	Digitalis glycosides
Eucharis spp.	Eucharist lily	Phenanthridine alkaloids
Euonymus spp.	Burning bush	Cardenolides, alkaloids
Euonymus spp.	Euonymus	Cardenolides (evonoside)
Eupatorium spp.	White snake root	Tremetol (ketone benzofuran)
Gloriosa superba	Glory lily	Colchicine
Helleborus spp.	Christmas rose, hellebore	Cardenolides, ranunculin
Hoya spp	Wax plant	Cardenolides
Hyacinthoides spp.	English bluebell	Cardenolides
Kalanchoe spp.	Kalanchoe	Cardenolides
Kalmia spp.	Laurel	Diterpenoids (Grayanotoxins)
Lobelia spp.	Cardinal flower	Pyridine alkaloids (lobeline)
Lyonia spp.	Staggerbush, maleberry	Diterpenoids (Grayanotoxins)
Nerium oleander	Oleander	Cardenolides
Nicotiana spp.	Tobacco	Piperidine alkaloids
Ornithogallum spp.	Star of Bethlehem	Cardenolides, saponins
Persea americanum	Avocado	Persin
Pieris spp.	Pieris	Diterpenoids (Grayanotoxins)

Genus	Common Name	Toxin
Rhododendron spp.	Rhododendron, azalea	Diterpenoids (Grayanotoxins)
Scilla spp.	Scilla, sqill	Bufadienolides
Taxus spp.	Yew	Alkaloids (taxine)
Thevetia spp.	Yellow oleander	Cardenolides

Plants Causing Neurologic Signs, Hallucinations

Aesculus spp.	Buckeye, horsechestnut	Esculin, escin
Albizia spp.	Mimosa tree	Vitamin B6 analogue
Asclepias spp.	Milkweed (narrow leaf)	Cardenolides
Atropa belladonna	Deadly nightshade	Tropane alkaloids (hyoscyamine)
Baptisia spp.	Indigo	Quinolizidine alkaloids (cytisine)
Blighia sapida	Akee	Amino acid Hypoglycin A,B
Brugmansia spp.	Angel's trumpet	Tropane alkaloids (hyoscyamine)
Brunfelsia spp.	Yesterday-today-tomorrow	Brunfelsamidine
Calotropis spp.	Crown flower	Cardenolides
Calycanthus spp.	Sweet shrub, allspice	Calycanthine
Cannabis sativa	Marijuana, hemp	Tetrahydrocannabinol
Catharanthus spp.	Periwinkle	Vincristine, reserpine
Chelidonium majus	Celandine	Berberine, chelidonine
Cicuta spp.	Water hemlock	Cicutoxin, cicutol
Clivia spp.	Clivia	Phenanthridine alkaloids (lycorine)
Colchicum spp.	Autumn crocus	Colchicine
Conium maculatum	European, spotted hemlock	Coniceine, coniine
Coriaria spp.	Coriaria	Coriamyrtin
Corydalis spp.	Fitweed, corydalis	Apomorphine, protoberberine
Crinum spp.	Crinum lily	Phenanthridine alkaloids (lycorine)
Cycas spp.	Cycad, sago palm	Glycosides
Datura spp.	Sacred datura, Jimson weed	Tropane alkaloids
Delphinium spp.	Larkspur	Diterpenoid alkaloids
Dicentra spp.	Bleeding heart	Isoquinoline alkaloids
Erythrina spp.	Coral bean	Alkaloids

continued

Table 2 Continued

Genus	Common Name	Toxin
Eucharis spp.	Eucharist lily	Phenanthridine alkaloids
Eupatorium spp.	White snake root	Tremetol (ketone benzofuran)
Gelsemium spp.	Carolina Jessamine	Gelsemine alkaloids
Gymnocladus spp.	Kentucky coffee tree	Gymnocladosapponins
Haemanthus spp.	Blood lily	Phenanthridine alkaloids
Humulus lupulus	Hops	Phenolic compounds
Lyonia spp.	Staggerbush, maleberry	Diterpenoids (Grayanotoxins)
Macadamia spp.	Macadamia nut	Unknown toxin
Melaleuca spp.	Melaleuca	Sesquiterpenes
Melia azadarach	Chinaberry	Triterpines (meliatoxins)
Nandina domestica	Heavenly bamboo	Cyanogenic glycosides, alkaloids
Nepeta cataria	Catnip	Monoterpene nepetalacetone
Nicotiana spp.	Tobacco	Piperidine alkaloids
Papaver spp.	Poppy	Phenanthrene alkaloids
Podophylum spp.	May apple	Podophyllotoxin
Solandra spp.	Chalice vine	Tropane glycoalkaloids
Solanum pseodocapsicum	Jerusalem cherry	Tropane glycoalkaloids
Sophora spp.	Mescal bean	Quinolizidine alkaloids (cytisine)
Sprekelia spp.	Maltese cross lily	Phenanthadine alkaloids
Theobroma cocoa	Cocoa	Methylxanthine alkaloids
Zamia spp.	Sago palm	Amino acids

Plants Causing Mydriasis

Atropa belladonna	Deadly nightshade	Tropane alkaloids (hyoscyamine)
Brugmansia spp.	Angel's trumpet	Tropane alkaloids (hyoscyamine)
Datura spp.	Sacred datura, Jimson weed	Tropane glycoalkaloids
Dracaena spp.	Dracaena, corn plant	Saponins

Genus	Common Name	Toxin
Nicotiana spp.	Tobacco	Piperidine alkaloids
Solandra spp.	Chalice vine	Tropane glycoalkaloids
Solanum pseodocapsicum	Jerusalem cherry	Tropane glycoalkaloids

Plants Causing Liver Disease

Agave lecheguilla	Agave	Sapogenins
Cestrum spp.	Cestrum, jasmine	Hepatotoxic glycosides
Cycas spp.	Cycad, sago palm	Glycosides
Heliotropium spp.	Heliotrope	Pyrrolizidine alkaloids
Juglans nigra	Black walnut	Juglone? (dogs experimentally)
Melaleuca spp.	Melaleuca, tea tree	Sesquiterpenes
Melia azadarach	Chinaberry	Triterpines (meliatoxins)
Senecio (Packera) spp.	Senecio, groundsel	Pyrrolizzidine alkaloids
Symphytum spp.	Comfrey	Pyrrolizidine alkaloids
Yucca spp.	Yucca	Saponins
Zamia spp.	Sago palm	Glycosides (cycasin)

Plants Causing Muscle Weakness, Tremors

Baptisia spp.	Indigo	Quinolizidine alkaloids (cytisine)
Brugmansia spp.	Angel's trumpet	Tropane alkaloids (hyoscyamine)
Brunfelsia spp.	Yesterday-today-tomorrow	Brunfelsamidine
Calycanthus spp.	Sweet shrub, allspice	Calycanthine
Cannabis sativa	Marijuana, hemp	Tetrahydrocannabinol
Corydalis spp.	Fitweed, corydalis	Apomorphine, protoberberine
Delphinium spp.	Larkspur	Diterpenoid alkaloids
Erythrina spp.	Coral bean	Alkaloids
Eupatorium spp.	White snake root	Tremetol (ketone benzofuran)
Humulus lupulus	Hops	Phenolic compounds
Macadamia spp.	Macadamia nut	Unknown toxin
Melaleuca spp.	Melaleuca, tea tree	Sesquiterpenes
Oxalis spp.	Oxalis, sorrel	Oxalates
Sophora spp.	Mescal bean	Quinolizidine alkaloids (cytisine)

continued

Table 2 Continued

Plants Causing Lameness

Genus	Common Name	Toxin
Juglans nigra	Black walnut	Juglone?
Cestrum diurnum	Day blooming jasmine	Vitamin D analogue

Plants That Cause Sudden Death

Eriobotrya spp.	Loquat	Cyanogenic glycosides
Gloriosa superba	Colchicine	Glory lily
Nandina domestica	Heavenly bamboo	Cyanogenic glycosides, alkaloids
Nerium oleander	Oleander	Cardenolides
Prunus spp.	Choke cherry	Cyanogenic glycosides
Taxus spp.	Yew	Alkaloids (taxine)
Thevetia spp.	Yellow oleander	Cardenolides

Plants Affecting the Skin

Ficus spp.	Fig	Proteolytic enzymes, phototoxin
Melaleuca spp.	Melaleuca, tea tree	Sesquiterpenes
Mimosa spp.	Sensitive plant	Mimosine
Schinus spp.	Pepper tree	Terpenoids

Table 3
Poisonous Fruits and Seeds of Common Garden Plants

Common Name	Genus/species	Toxin Class	Clinical Effect
Acorn	Quercus spp.	Gallotannins	Kidney failure
Akee	Blighia sapida	Amino acid Hypoglycin	Gastroenteritis
Arum	Calla spp.	Oxalates	Kidney failure
Asparagus	Asparagus	Saponins	Gastroenteritis
Balsam pear	Momordica charantia	Cucurbitane terpenoids	Gastroenteritis
Baneberry	Actaea spp.	Ranunculin, protoanemonins	Gastroenteritis
Belladonna	Atropa belladonna	Tropane alkaloids	Respiratory failure
Bitter sweet	Celastrus scandens	Cardenolides	Gastroenteritis
Black henbane	Hyoscyamus niger	Tropane alkaloids	Respiratory failure
Blue cohosh	Caulophyllum thalictroides	Glycoside	Gastroenteritis
Box	Buxus spp.	Steroid alkaloids	Gastroenteritis
Buckeye	Aesculus spp.	Glycosides - esculin, escin	Vomiting, seizures
Buckthorn	Rhamnus spp.	Cascarocide glycosides	Purgation
Buddist pine	Podocarpus spp.	(Unknown)	Gastroenteritis
Calabar bean	Physostigma venenosa	Physostigmine	Salivation, miosis
Candle nut	Aleurities spp.	Diterpene alkaloids	Gastroenteritis
Castor bean	Ricinus communis	Lectins (toxalbumins)	Gastroenteritis
Chili peppers	Capsicum spp.	Capsaicin	Gastroenteritis
Clivia	Clivia spp.	Oxalates	Kidney failure
Chinaberry	Melia azadarach	Triterpenes – meliatoxins	Gastroenteritis
Chinese lantern	Physalis spp.	Solanine glycoalkaloids	Respiratory failure
Cocklebur	Xanthium spp.	Carboxyactractyl-oside	Liver failure
Coontie	Zamia spp.	Glycosides – cycasin	Liver failure

Table 3 Continued

Common Name	Genus/species	Toxin Class	Clinical Effect
Coral tree	Erythrina spp.	Isoquinolone alkaloids	Muscle weakness
Coyotillo	Karwinskia humboldtiana	Glycosides	Gastroenteritis
Cycad	Cycas spp.	Glycosides – cycasin	Liver disease
Daphne	Daphne mezereum	Diterpene alkaloids	Gastroenteritis
Deadly nightshade	Atropa belladonna	Tropane glycoalkaloids	Respiratory failure
English ivy	Hedera spp.	Terpenoids	Gastroenteritis
Heavenly bamboo	Nandina domestica	Cyanogenic glycosides	Respiratory failure
Holly	Ilex spp.	Methylxanthines, saponins	Gastroenteritis
Honey suckle	Lonicera spp.	Terpenoids	Gastroenteritis
Jack in the pulpit	Arisaema spp.	Oxalates	Kidney failure
Jerusalem cherry	Solanum pseudocapsicum	Tropane alkaloids	Respiratory failure
Jessamine	Cestrum spp.	Vitamin D analogue	Tissue calcification
Jimson weed	Datura spp.	Tropane alkaloids	Respiratory failure
Kentucky coffee tree	Gymnocladus dioiacea	Glycosides	Gastroenteritis
Lantana	Lantana spp.	Triterpenoids, lantadenes	Cholecystitis
Lily of the valley	Convallaria majalis	Cardenolides	Cardiac failure
Loquat	Eryiobotria japonica	Cyanogenic glycosides	Respiratory failure
Manchineel	Hippomane mancinella	Diterpene alkaloids	Gastroenteritis
May apple	Podophyllum	Podophyllotoxin	Gastroenteritis
Mescal bean	Sophora secundiflora	Quinolizidine alkaloids	Neurotoxic
Mistletoe	Phoradendron spp.	Glycoprotein lectins	Gastroenteritis
Moonseed	Menispermum canadense	Alkaloids	Gastroenteritis, seizures
Morning glory	Ipomoea spp.	Indole alkaloids	Hallucinations

Common Name	Genus/species	Toxin Class	Clinical Effect
Pepper tree	*Schinus* spp.	Terpenes	Gastroenteritis
Physic, purge nut	*Jatropha* spp.	Diterpenoids	Gastroenteritis
Pigeon berry	*Duranta repens*	Monoterpenoids	Gastroenteritis
Pokeberry	*Phytolacca americana*	Saponins, oxalates	Gastroenteritis
Poison wood	*Metopium toxiferum*	Catechols	Dermatitis
Privet	*Ligustrum* spp.	Terpenoids	Gastroenteritis
Rattlepod	*Crotolaria* spp.	Pyrrolizidine alkaloids	Liver failure
Rosary pea	*Abrus precatorius*	Lectins (toxalbumins)	Gastroenteritis
Scotch broom	*Cytissus* spp.	Quinolizidine alkaloids	Neurotoxic, emesis
Snow berry	*Symphoricarpos* spp.	Alkaloids	Gastroenteritis
Soap berry	*Sapindus* spp.	Saponins	Gastroenteritis
Strawberry bush	*Euonymous* spp.	Cardenolides	Gastroenteritis
Sweet Pea	*Lathyrus* spp.	Lathyrogenic aminoacids	Neural degeneration
Virginia creeper	*Parthenocissus quinquefolia*	Oxalates	Kidney failure
Woody nightshade	*Solanum dulcamara*	Tropane glycoalkaloids	Gastroenteritis
Yellow oleander	*Thevetia peruviana*	Cardenolides	Cardiac failure
Yew	*Taxus* spp.	Taxine	Cardiac failure

Abrus precatorius
Family: Fabaceae (Leguminosae)

Common Names
Rosary or prayer pea, crab's eye, black-eyed
Susan, coral bead plant, gidee-gidee,
Indian bead guinea pea, jequirity bean,
love pea, lucky bean, jumbee beads,
peonia, pukiawe-lei, Seminole bead, wild
licorice, or licorice vine

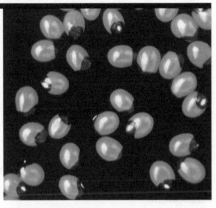

Figure 1 *Abrus* seeds

Plant Description
Originating in India, Abrus has become
widely distributed in tropical areas
including Florida, Carribean, and the
Hawaii islands. It is a woody, perennial,
slender vine, that grows over other
vegetation. The leaves are pinnately
compound, with 8-15 pairs of leaflets. The
lavender-pink to pale red pea-like flowers
are produced as racemes from leaf axils.
The leguminous pods are flat, pubescent,
beaked, and when ripe, unfurl to reveal the
characteristic attractive scarlet red peas with
a black end (Figures 1 and 2).

Toxic Principle and Mechanism of Action
Only the seeds of the rosary pea are toxic
as they contain potent lectins
(toxalbumins) called abrin I and II that are
toxic to all animals including humans.
Abrin is very similar to ricin, the lectin
found in castor beans, and consists of 2

Figure 2 *Abrus* vine and seed pod.

polypeptide chains (A and B), cross-linked by a disulfide bond that is a potent
ribosomal inhibitor.[1,2] The B chain binds to carbohydrate receptors on cell surfaces,
facilitating the entry of the A chain into the cell where it inhibits initiation and
elongation of peptides within ribosomes. Rapidly growing and dividing cells such as
those of the intestines are most severely affected. The lethal dose (LD_{50}) of abrin is
in the range of 0.1-0.2 micrograms/kg body weight.[2] Each gram of seed contains
approximately 0.5mg of abrin.[2]

Risk Assessment
Most poisoning by *Abrus precatorius* is reported in children who eat the attractive
peas.[4] However, all animals including dogs, poultry and other birds, horses, pigs,
and ruminants are susceptible if the hard seeds are well chewed before being
swallowed.[5] Intact seeds when swallowed pass through the digestive system

without exposing the animal to the lectins contained within the seed. The greatest risk to animals are the seeds which are often collected and brought into the household and become accessible to cats, dogs, and caged birds.

Clinical Signs

Abdominal pain, bloat, and hemorrhagic diarrhea develop up to a day after the ingestion of a toxic dose of abrin. Excessive salivation, vomiting, and diarrhea can lead to severe dehydration, hypovolemic shock, and death.

Post mortem examination frequently will show reddening, hemorrhages and ulceration of the gastrointestinal tract. Other organs may show similar gross lesions. Histologically there is hepatic and renal degeneration, vascular congestion, hemorrhaging, and ulceration of the mucosal surfaces.

Treatment

Aggressive intravenous fluid and electrolyte therapy may be necessary to counteract severe dehydration. Activated charcoal orally is indicated to reduce further absorption of abrin from the digestive tract

References

1. Hedge R, Maiti TK, Poddler SK: Purification and characterization of three toxins and two agglutinins from *Abrus precatorius* seed by using lactymal-sepharose affinity chromatography. Anal Biochem 1991, 194: 101-109.

2. Olsnes S, et al: Mecanisms of action of 2 toxic lectins, abrin and ricin. Nature 1974, 249: 627-631.

3. Burrows GE, Tyrl RJ: *Abrus* in Poisonous Plants of North America. (Eds) Burrows GE and Tyrl RJ. Iowa State University Press, Ames. 2001. 501-503.

4. Davis J: *Abrus precatorius* (rosary pea), the most common lethal poisonous plant. J Florida Med Assoc 1978, 65: 189-191.

5. Knight MW, Dorman DC: Selected poisonous plant concerns in small animals. Vet Med 1997, 92: 260.

Family: Euphorbiaceae

Common Names

Copper leaf, Fijean fire-plant, red-hot cat-tail, chenille plant, Joseph's coat, three-seeded mercury, Phillipine medusa

Plant Description

Over 400 species of this plant occur in the tropical areas of the world. Twenty species occur in North America, one of which *A. virginica* (Virginia copperleaf), has been associated with poisoning in humans (Figures 3 and 4).

Figure 3 *Acalypha hispida*

Toxic Principle and Mechanism of Action

Diterpene esters are present in the plant are thought to be responsible for the gastro intestinal irritation associated with ingestion of the plant. The latex from the plant can cause skin irritation. Anthocyanins have been isolated from the plant.[1] The cyanogenic glycoside acalyphin has been identified in *Acalypha indica*, and is toxic to livestock.[2]

Risk Assessment

There are no reported cases of poisoning in animals involving *Acalypha* species. However, some species may be associated with gastroenteritis and dermatitis in people who handle the plant.[3] It is a low risk plant.

Clinical Signs

Nonspecific gastroenteritis and dermatitis **Figure 4** *Acalypha wilkesiana* Copper leaf
can be expected in persons handling the plant.

References

1. Reiersen B, Kiremire BT, Byamukama R, Andersen MOM: Anthocyanins acylated with gallic acid from chenille plant, *Acalypha hispida*. Phytochemistry. 2003, 64:867-871.
2. Nahrstedt A, Kant JD, Wray V: Acalyphin, a cyanogenic glucoside from *Acalypha indica*. Phytochemistry 1982, 21: 101-105.
3. Spoerke DG, Smolinske SC: Toxicity of Houseplants. CRC Press, Boca Raton, Florida. 2003, pp 62.

Acer

Family: Aceraceae

Figure 5 Silver under side to leaves

Figure 6 Red maple Fall foliage

Common Name
Maples, red or swamp maple *(Acer rubrum)*, silver maple *(A. saccharinum)*

Plant Description
A genus of about 150 species of deciduous shrubs to large trees native to North and central America, Europe, Asia, and North Africa. Leaves are palmate, opposite, shallow to deeply lobed, although some species have no lobes and are pinnate. In the Fall the leaves turn various shades of yellow to deep red. The inflorescences are produced before the leaves and are pendulous, panicles or racemes terminally or in leaf axils. Flowers are small, with 5 greenish yellow or red sepals and 5 red petals. Some may have 4-6 perianth parts. The fruits (samaras) are characteristically winged, paired, and indehiscent. (Figures 5 and 6)

Toxic Principle and Mechanism of Action
The consumption of the wilted or partially dried leaves of the red maple *(Acer rubrum)* is known to cause acute destruction of equine red blood cells.[1-4] Fresh, green maple leaves do not appear to be toxic. The toxin responsible for the oxidation of the hemoglobin in the red blood cells has not been fully determined. However, extracts obtained from various maples including *(A. rubrum, A. saccharum, A. saccharinum)* when incubated with horse red blood cells increased the formation of methemoglobin.[5] The '*Acer* fraction' causing methemoglobinemia contained gallic acid, and other oxidants.[5] Other compounds such as phenols, amines and nitro-compounds produce methemoglobinemia and may also be present in maple leaves.[6] Although reported cases of poisoning have been associated with red maple poisoning, it is likely that other maples including the silver maple and the sugar maple are capable of similar poisoning.[5] Horses are most frequently poisoned by red maple leaves, but all *Equidae* species including zebras are likely susceptible.[2,3,7] Alpacas are also susceptible to red maple poisoning.[8]

Risk Assessment

Red maples and its numerous cultivars ('October glory', 'Red sunset', 'Bowhall', 'Scanlon', 'Schlesingeri') are often grown for their attractive red Fall foliage. Until proven otherwise red maple and its cultivars should be considered toxic to equids including horses, mules, and zebras. It is inadviseable to plant red maples in or around horse pastures or enclosures. Because other *Acer* species (Sugar maple, silver leaf maple) have been shown to contain oxidants, similar precautions should be considered when selecting these species where horses are likely to be exposed to the leaves.[5]

Clinical Signs

Several days after eating wilted maple leaves, horses become depressed, anorexic and develop dark, red-brown colored urine.[1-4] If a large quantity of maple leaves are consumed at one time, a horse may become very depressed, cyanotic and die within a 24 hour period. Typically, the signs progress in severity with the animal becoming more depressed, weaker, with increased respiratory and heart rates, and cyanotic. Anemia develops rapidly as the damaged red blood cells are destroyed. The hematocrit may drop to 10%, and numerous Heinz bodies may be seen in the red blood cells.[1,2] Heinz bodies may persist for several weeks. Methemoglobin levels are usually markedly elevated. Liver enzymes, bilirubin, creatinine levels may also be elevated. Glutathione levels in erythrocytes are markedly decreased. Abortion can occur in pregnant mares consuming red maple leaves.[9] Death may occur anywhere up to a week after the ingestion of the maple leaves and is generally due to the combined effects of severe anemia and renal failure.[1-4]

Treatment

Since the signs of red maple toxicity are often not recognized until several days after the consumption of the leaves, oxidant injury is quite advanced and a guarded prognosis is warranted. Affected horses should not be stressed or excited because of the severe anemia and methemoglobinemia affecting oxygen transportation. Intravenous fluids and blood transfusions may be necessary to correct the anemia and support renal function. Large doses of ascorbic acid orally (125mg/kg body weight) and intravenously (30 mg/kg body weight) every 12 hours in conjunction with anti-inflammatory drugs and intravenous fluids have been used successfully to counter the oxidant effects of the maple leaves.[10] Activated charcoal administered via stomach may help to prevent further absorption of the toxins from the gastrointestinal tract. Methylene blue should not be used in horses to treat the methemoglobinemia as it enhances the oxidant injury to the erythrocytes.

References

1. Tennant B, et al: Acute hemolytic anemia, methemoglobinemia, and heinz body formation associated with ingestion of red maple leaves by horses. J Am Vet Med Assoc 1981, 179: 143-150.
2. George LW, Divers TJ, Mahaffey EA, Suarez MJ: Heinz body anemia and methemoglobinemia in ponies given red maple *(Acer rubrum L.)* leaves. Vet Path 1982, 19: 521-533.

3. Divers TJ, George LW, George JW: Hemolytic anemia in horses after the ingestion of red maple leaves. J Am Vet Med Assoc1982, 180: 300-302.
4. Plumlee KH: Red maple toxicity in a horse. Vet Hum Toxicol 191, 33: 66-67.
5. Boyer JD, Breeden DC, Brown DL: Isolation, identification, and characterization of compounds from *Acer rubrum* capable of oxidizing equine erythrocytes. Am J Vet Res 2002, 63: 604-610.
6. Smith RP: Toxic responses of the blood. In Casarett & Doull's Toxicology, 4th ed. Amdur MO, Doull J, Klassen CD eds. Pergamon Press, New York 1993, 257-281.
7. Weber M, Miller RE: Presumptive red maple (Acer rubrum) toxicosis in Grevy's zebra *(Equus grevyi)*. J ZooWildlife Med 1997, 28: 105-108.
8. DeWitt SF, Bedenice D, Mazan MR: Hemolysis and Heinz body formation associated with ingestion of red maple leaves in two alpacas. J Am Vet Med Assoc 2004, 225: 578-583.
9. Stair EL, Edwards WC, Burrows GE, Torbeck K: Suspected red maple *(Acer rubrum)* toxicosis with abortion in two Percheron mares. Vet Human Toxicol 1993, 35: 229-230.
10. McConnico RS, Brownie CF: The use of ascorbic acid in the treatment of 2 cases of red maple *(Acer rubrum)*-poisoned horses. Cornell Vet 1992, 82: 293-300.

Achillea

Family: Asteraceae

Figure 7 *Achillea* 'Coronation gold'

Common Name
Yarrow, milfoil, sneezewort

Plant Description
Consisting of about 85 species, mostly from Europe and Asia, with a few species in North America, the *Achillea* are commonly grown as garden plants for their multicolored long-lasting, erect, flat flower heads consisting of many small daisy-like flowers, red, white, yellow, or red in color (Figures 7-9). The perennials are often aromatic with leaves arising from a spreading rhizome. The leaves are evenly spaced along the stems, lanceolate in shape, with finally divided segments.

Toxic Principle and Mechanism of Action
Achillea species contain a variety of glycoalkaloids, monoterpenes, and sesquiterpene lactones that act as gastrointestinal irritants.

Risk Assessment
There are no reported cases of poisoning in animals involving *Achillea* species, except in sheep that experienced digestive disturbances after eating the plant. The aromatic

yarrows are unlikely to be consumed by pets, but if they are, gastrointestinal disturbances should be anticipated.

Clinical Signs
Non specific gastroenteritis with vomiting and diarrhea. Treatment is rarely necessary once the source of the plant is removed.

References
1. Millspaugh CF: American medicinal plants. Dover, 1974, 305-377.
2. Hanlidou E, Kokkalou E, Kokkini S: volatile constituents of Achillea grandifolia. Plata Med 1992, 58:105-107.

Figure 8 *Achillea* 'Cerise queen'

Figure 9 *Achillea milleforme*

Acokanthera
Family: Apocynaceae

Common Name
Bushman's poison bush

Plant Description
Acokanthera species are evergreen shrubs with glossy green, leathery, elliptical, opposite leaves that terminate in sharp points. The tubular flowers are produced in clusters from the leaf axles. The white flowers are fragrant and pinkish on the outside. The fruits are droops containing a milky spongy flesh and two seeds. *Acokanthera* species, contain a milky sap that is poisonous. (Figures 10-12).

Figure 10 *Acokanthera oblongifolia*

Figure 11 *Acokanthera oblongifolia fruit*

Figure 12 *Acokanthera oppositifolia*
Photographs courtesy of E. Vinter University of Preforia, Onderstepoort, South Africa

These attractive flowering shrubs are native to Africa and have been imported into tropical areas such as Florida. Variegated forms of the plant have been developed.

Toxic Principle and Mechanism of Action

Acokanthera species contain several potent cardiac glycosides such as ouabain and acokantherin.[1] Poisoning has been reported in cattle, sheep, goats, donkeys, ostriches, and humans.[1] The milky sap has been used as an arrow poison in southern Africa.[2]

Risk Assessment

Although a relatively uncommon garden plant in North America, it's potent fast acting cardiotoxins warrant caution to ensure household pets and children do not have access to the plant. All parts of the plant are poisonous.

Clinical Signs

Animals consuming any part of the plant can be anticipated to develop clinical signs ranging from cardiac arrhythmias, heart block, to death within 1-2 hours after eating the plant.

Specific Treatment

Considering the nature of the toxins present in *Acokanthera* species, treatment would be as for digitalis poisoning (See Digitalis).

Reference

1. Kellerman TS, Coetzer JAW, Naude TW (Eds). Plant Poisonings and Mycotoxicoses of Livestock in Southern Africa. Oxford University Press, Cape Town, 1990, pp 86.
2. Watt JM, Breyer-Brandwijk Mg: The Medicinal and poisonous plants of southern and eastern Africa. E&S Livingston, Edinburgh 1962, pp62-78.

Aconitum

Family: Ranunculaceae

Common Name

Monkshood, wolfsbane, aconite, helmet flower, friar's cap

Plant Description

A common garden plant, with about 100 species of this genus found world wide. There are 5 native species in North America *(Aconitum columbianum, A. napellus, A. delphinifolium, A. maximum, A. reclinatum, A. uncinatum)*[1]

Perennial herbs, growing from a tuberous root system, *Aconitum* species are erect (up to 6 feet in height), or sprawling plants with alternate, palmately-lobed, some with markedly cleft leaves, and flowers produced in terminal racemes. Flowers have 5 sepals, the uppermost one being distinctly hood-shaped, often with a prominent beak. Petals are small (2 or 5), contained within the hooded sepal. Flower color is generally a deep blue or purple, but can range from white to a yellow-green (Figures 13 and 14). Fruits are oblong, beaked follicles with numerous seeds.

Figure 13 *Aconitum columbianum*

Figure 14 Monkshood flowers

Toxic Principle and Mechanism of Action

Aconitum species, and all parts of the plant contain highly toxic diterpenoid alkaloids, the most toxic of which is aconitine.[2] These diterpenoid alkaloids suppress the inactivation of voltage-dependent Na^+ channels by binding to neurotoxin binding site 2 of the alpha - subunit of the channel protein.[3] This results in depolarization of nerve and muscle cells, causing cardiac arrhythmias and muscle weakness. Aconitine also acts on the central nervous system affecting the adrenergic and cholinergic systems.[3] Similar toxic diterpenoid alkaloids are found in larkspurs (*Delphinium* species). Aconitine is readily absorbed through the skin and mucous membranes.

Risk Assessment

Although it is rarely reported as a problem to dogs and cats, *Aconitum* is quite frequently a cause of poisoning in humans, especially where herbal preparations from monkshood have been used for medicinal purposes.[3-6] Monkshood is poisonous to cattle, and other animals that might eat the plant.[7] Recognition of the fact that monkshood is highly toxic warrants caution in planting it in the garden where children and pets may have opportunity to eat it.

Clinical Signs

The primary signs of monkshood poisoning are those resulting from its effects on the autonomic nervous system: salivation, vomiting, diarrhea, heart irregularity and fibrillation, muscle tremors and weakness, respiratory difficulty and death in severe cases.

Specific Treatment

Because of the similarity of the aconitine alkaloids to those found in *Delphinium* species, similar treatment methods can be applied as for larkspur poisoning. To help reverse the neuromuscular blocking effects of the alkaloids in monkshood poisoning, physostigmine may be effective. Cardiac monitoring and other symptomatic treatment should be provided.

References

1. Brink DE, Woods JA: *Aconitum.* In Flora of North America, Vol 3.Flora of North America Editorial Committee (Eds). Oxford University Press, New York, 1997, 191-195.
2. Jaycyno JM: Chemistry and toxicology of the diterpenoid alkaoids. In Chemistry and Toxicology of Diverse Classes of Alkaloids. Blum MS (Ed) Alaken, Fort Collins, Colorado. 1996, 301-336.
3. Ameri A: The effects of Aconitum alkaloids on the central nervous system. Progress in Neurobiology. 1998, 56: 211-235.
4. Chan TYK, Tomlinson B, Tse LKK, Chan JCN, Chan WWM: Aconitine poisoning due to Chinese herbal medicines: A review. Vet Human Toxicol. 1994, 36: 452-455.
5. Pui-Hay But P, Tai Yau-Ting, Young K: Three fatal cases of herbal aconite poisoning.Vet Human Toxicol 1994, 36: 212-216.
6. Chan TYK: Aconitine poisoning: A global perspective. Vet Human Toxicol 1994, 36: 326-328.
7. Puschner B, Booth MC, Tor ER, Odermatt A: Diterpenoid Alkaloid Toxicosis in Cattle in the Swiss Alps. Vet Human Toxicol. 2002, 44: 8-10.

Family: Ranunculaceae

Common Name

Red bane berry, white bane berry *(Actaea alba)*, Dolls eye snakeberry, necklace weed, red or white cohosh. *(Actaea erythrocarpa, A. pachypoda*, are European species)* The plant known as white cohosh *(A. pachypoda)* should not be confused with blue cohosh or squaw root *(Caulophyllum thalictroides)*, or black cohosh *(Cimicifuga racemosa)*, woodland plants of northern temperate areas, containing various toxic saponins, alkaloids, and methylcystine. Blue and black cohosh have been used in herbal medicine as abortifacients and for menstrual problems in women, but not without problems of toxicity including congestive heart failure.[1-3]

Figure 15 *Actaea alba*

Plant Description

The 8 known species of *Actaea* are woody perennial plants of Europe and North America, occurring mostly in moist woodland areas. Growing to 3 ft. in height, the plant grows from a root crown and has large compound, opposite leaves with distinct sharp teeth, and strong veining. Flowers are produced in short, terminal spikes, individual flowers small white with many white stamens.

Figure 16 *Actaea rubra* flower and fruits

The fruits are either red, white, or black berries *(A. spicata)*, often on a stalk of contrasting color (Figures 15 and 16).

Toxic Principle and Mechanism of Action

The attractive red or white berries and the roots contain the glycoside ranunculin which is hydrolysed to protoanemonin when eaten. Protoanemonin is an irritating compound found in many of the Buttercup family (Ranunculaceae).[4,5]

Risk Assessment

Being mildly toxic, poisoning occurs primarily in children who eat the attractive red or white berries. However, the berries have the potential for being eaten by household pets, should they be brought into the household.

Clinical Signs

Depending on the quantities of berries consumed, excessive salivation, vomiting, and diarrhea may result. Fatalities are unlikely. Protoanemonin can be secreted in the milk of lactating animals and will impart a bitter taste to the milk. The signs of gastrointestinal irritation are usually self-limiting because the quantity of the berries consumed is limited by their bitter taste.

References

1. Rao RB. Hoffman RS. Nicotinic toxicity from tincture of blue cohosh *(Caulophyllum thalictroides)* used as an abortifacient. Vet Human Toxicol 2002, 44: 221-222.
2. Jones TK. Lawson BM. Profound neonatal congestive heart failure caused by maternal consumption of blue cohosh herbal medication. J Pediatrics 1998, 132: 550-552.
3. Li W, Sun Y, Liang W, Fitzloff JF, van Breemen RB: Identification of caffeic acid derivatives in *Actea racemosa (Cimicifuga racemosa,* black cohosh) by liquid chromatography/tandem mass spectrometry. Rapid Communications in Mass Spectrometry, 2003, 17: 978-982.
4. Burrows GE, Tyrl RJ: Toxic Plants of North America. Iowa State University Press, Ames, 2001, 1004-1005.
5. Lampe KF, McCann MA: AMA Handbook of Poisonous and Injurious Plants. Am Med Assoc, Chicago, Illinois, 1985, pp 21-22.

Adenium obesum

Family: Apocynaceae

Figure 17 *Adenum* flowers

Common Name

Desert rose, mock azalea, desert azalea, impala lily, kudu lily

Plant Description

Adenium is a genus with one variable species originating from the desert areas of the Arabian peninsula extending South into Africa. It has characteristic swollen trunks and brilliant funnel-shaped flowers. Leaves are glossy, lanceolate, and produced in whorls at the tips of the branches. The plant is usually leafless when in flower. Flower color can vary from pale pink to deep red in the tropical dry season. Varieties of this species are becoming popular potted houseplants and garden plants in hot climactic zones (Figures 17 and 18).

Toxic Principle and Mechanism of Action

Some 30 cardiac glycosides with similarity to the toxic glycosides found in oleander and *Acokanthera* species, have been isolated from the plant. The plant has been used as a source of arrow poison in Africa.[1,2]

Risk Assessment

Although there are no reported cases of poisoning in animals, this plant is sufficiently toxic to warrant caution in introducing it to a household where the plant may be consumed by animals.

Clinical Signs

If a toxic dose of the plant is consumed a variety of cardiac arrhythmias, heart block, and death could be anticipated. Treatment would be as for oleander poisoning, or other cardiac glycoside toxicities.

References

1. Yamauchi T, Abe F: Study on the constituents of Adenium.1. Cardiac glycosides and pregnanes from *Adenium obesum*. Chem Pharm Bull 1990, 38: 669-672.
2. Kellerman TS, Coetzer JAW, Naude TW: (Eds). Plant Poisonings and Mycotoxicoses of Livestock in Southern Africa. Oxford University Press, Cape Town, 1990, pp 86.

Figure 18 *Adenium obesum*

Adonis

Family: Ranunculaceae

Common Name

Pheasant's eye, Adonis, yellow ox-eye, flos adonis

Plant Description

The genus *Adonis* consists of at least 20 species of annuals and perennials originating primarily in Europe and cooler parts of Asia. Three introduced species are common in North America: *Adonis aestivalis* (red flower), *A. annua* (red flower with black center), *A. amurensis* (yellow flowers).

Figure 19 *Adonis aestivalis*

Closely related to the anemone and other members of the Buttercup (Ranunculus) family, *Adonis* are either annuals or perennials, arising from a tap root or rhizome, and with simple alternate finely divided pinnate leaves. Depending on the species, some *Adonis* grow up to 2 ft. and produce single solitary flowers terminally on the branches. Flowers have 5-8 greenish sepals and from 5-20 colorful petals, ranging in color from yellow to red, and occasionally white. Some species have a distinct black

Figure 20 *Adonis amurensis*

spot at the base of the petals (Figures 19 and 20). Fruits are beaked globular achenes.

All three of the species occurring in North America were originally introduced as garden plants which have escaped from cultivation. *Adonis aestivalis* in particular, has in some areas escaped to invade alfalfa fields and cause poisoning in horses fed hay containing the plant.[1,2]

Toxic Principle and Mechanism of Action

Adonis species contain numerous cardenolides, including strophanthidin glycosides that are cardiotoxic.[3] Being a member of the Buttercup family, and considering that the clinical signs produced from eating *Adonis* are those of gastroenteritis, it is likely that the plants also contains ranunculin, which is hydrolyzed to the irritant protoanemonin once chewed and swallowed.

Risk Assessment

The *Adonis* species generally have a bitter taste and are not palatable. However as the species become popular garden plants, there is increased risk that household pets may be exposed to them. *Adonis aestivalis* has demonstrated that it can readily escaped from cultivation and become a problematic weed in hay feels thereby posing a risk to horses and other animals fed hay containing the plant.

Clinical Signs

The clinical signs of *Adonis* poisoning are primarily those of a gastroenteritis, resulting in vomiting and diarrhea. The diarrhea can be severe, causing dehydration and colic that will warrant appropriate supportive and symptomatic treatment. Myocardial necrosis can result in death in horses eating hay containing *Adonis.*[2]

References

1. Everist SL: Poisonous Plants of Australia. 2nd ed. Angus and Robertson, Sydney, Australia, 1981, 596-603.
2. Woods LW et al: Myocardial degeneration and death in three horses with ingestion of grass hay containing summer Adonis *(Adonis aestivalis)* Am Assoc Vet Lab Diagn. 46th Ann Conf Proceedings, San Diego 2003, pp124.
3. Joubert JPJ, Colin: Cardiac glycosides. In Toxicants of Plant Origin, vol 2, Glycosides, PR Cheeke (Ed). CRC Press, Boca Raton, Florida, 1989, 61-96.

Aesculus

Family: Hippocastanaceae

Common Name
Horse chestnut, buckeye, fish poison, marronnier. California buckeye *(A. californica)*, Ohio buckeye *(A. glabra)*, red buckeye *(A. pavia)*, horsechestnut *(A. hippocastanu)*

Plant Description
A genus of some 20 species of trees and shrubs, *Aesculus* are frequently planted as ornamental trees for their attractive flowers, leaves, and fruits. At least half of the 20 known species grow in North America. Extensive hybridization has occurred amongst the species. These deciduous trees generally prefer moist soil conditions and can develop into large specimen trees. Trunks have gray to brown smooth or rough bark, the characteristic large palmate leaves have 5-11 leaflets. Inflorescences are produced terminally on branches, either before or after leads are produced. Flower color is typically white, creamy-yellow, or red (Figures 21 and 22). Fruits are smooth or warty capsules that break open in the fall to release one to three dark brown shiny seeds that have a characteristic lighter scar on one side (Figures 23 and 24).

Figure 21 *Aesculus carnea*

Figure 22 *Aesculus carnea*

Toxic Principle and Mechanism of Action
Aesulus species contain a variety of toxins in the buds, leaves, and seeds including esculin (a coumarin glycoside), escin (complex saponin) and frangula (an anthraquinone).[1,2] These compounds, once hydrolyzed, become toxic and through their irritant effects on the gastrointestinal tract can induce colic and diarrhea. Esculin is known to be cytotoxic and may account for the neurologic signs seen in some cases of poisoning.

Risk Assessment
Horse chestnuts or buckeyes (conkers) are produced in large numbers in the fall and the attractive brown seeds are commonly collected and brought into households, where they become accessible to pets. The seeds are a potential source of poisoning to animals that eat them. Most poisoning reported in animals has occurred in livestock.[3,4] The edible chestnut (*Castanea* species) is not related to the *Aesculus* species.

Figure 23 Horse chestnuts

Clinical Signs

Gastrointestinal signs generally predominate, and consist of vomiting and diarrhea shortly after ingestion of the leaves or seeds. Colic is the principal clinical sign seen in horses. Depending on the quantity and frequency of *Aesculus* consumed, neurologic signs, including excitement and ataxia, muscle twitching, a stiff gait, seizures, recumbency, and death may occur in severe cases.

Treatment

Symptomatic treatment is indicated as there is no specific antidote. Inducing vomiting, if the animal is known to have eaten the seeds within the past two hours, is beneficial. Activated charcoal given via stomach tube with a laxative may help to inactivate the toxins and remove them from the gastrointestinal tract. Intravenous fluid therapy may be indicated in cases of severe dehydration. Animals showing neurologic signs and seizures may be treated with diazepam.

References

1. Yoshikawa M et al: Bioactive glycosides and saponins. III. Horse chestnut. Chem Pharm Bull 1996, 44: 1454-1464.

Figure 24 *Aesculus hippocastanum*

2. Yoshikawa M et al: Bioactive saponins and glycosides. 3. Horse chestnut. The structures, inhibitory effects on ethanol absorption, and hypoglycaemic activity of the escins Ia,Ib,Iia, Iib, and IIIa from the seeds of *Aesculus hippocastanum* L. Chem Pharm Bulletin 1996, 44: 1454-1464.

3. Kornheiser KM: Buckeye poisoning in cattle. Vet Med/Sm Anim Clin 1983, 78 :769-770.

4. Casteel SW, Wagstaff DJ: *Aesulus glabra* poisoning in cattle. Vet Hum Toxicol 1983, 34: 55- .

Agapanthus africanus

Family: Liliaceae

Common Name
African blue lily, agapanthus, lily of the Nile

Plant Description
Native to southern Africa, the agapanthus lily is a vigorous perennial in mild climates, and has gained popularity as a potted houseplant in cold climates. Long, arching, strap-shaped leaves are produced from short rhizomes with fleshy roots. Flowers are produced in umbels on tall stems and come in various shades of blue, and occasionally white. Numerous cultivars of this lily are available commercially (Figures 25 and 26).

Figure 25 *Agapanthus africanus*

Toxic Principle and Mechanism of Action
No specific toxin has been isolated from the agapanthus lily although it is reported to contain saponins and sapogenins.[1] *Agapanthus* species have been used in phytomedicine in some parts of Africa to cause abortion and as an aphrodisiac.[1] Aqueous extracts from the plant have been shown to have pharmacologic effects upon the uterus causing contractions, possibly due to the production of prostaglandins.[2]

Figure 26 *Agapanthus* flower

Risk Assessment
There are no known cases of agapanthus poisoning in animals. However, the plant is increasingly more common in households with pets, and therefore warrants recognition as a potentially toxic plant.

References
1. Watt JM, Breyer-Brandwijk MG: The medicinal and poisonous plants of southern and eastern Africa. E & S Livingston, Edinburgh, 1962, pp 669-670.
2. Veale DJ, Havlik I, Oliver DW, Dekker TG: Pharmacological effects of *Agapanthus africanus* on the isolated rat uterus. J Ethnopharmacology. 1999, 66: 257-262.

Agave

Family: Agavaceae

Figure 27 *Agave lecheguilla*

Figure 28 *Agave parryi*

Common Name
Agave, century plant, American aloe, sisal, maguey

Plant Description
Primarily plants of the desert areas of the southwestern United States, Mexico, and the Caribbean islands, many of the Agave (300 species) have become popular in xeriscape landscaping, and occasionally as houseplants. These long-lived perennial plants vary considerably in size, the leaves ranging from 1-2 in. to 4-5 ft. in length. The leaves are fleshy, smooth, alternate, generally sword-shaped, often with sharply toothed leaf edges and a terminal spine (Figures 27 and 28). Small species may take up to 5 years to produce a flower spike while some of the large species may not flower for up to 50 years. Each plant only flowers once in its lifetime, and after flowering dies, but produces offsets that maintain the plant. The numerous flowers are produced on tall spikes or racemes. Fruits are capsules or berries containing numerous seeds.

Toxic Principle and Mechanism of Action
The only species of Agave known to cause poisoning in animals is *A. lecheguilla* (lecheguilla) and its toxicity is due to the presence of hepatotoxic sapogenins.[1] These compounds affect the liver's ability to excrete the photodynamic compound phylloerythrin via the biliary system. Cholestasis appears to result from the accumulation of calcium sapogenin salts in the bilary ducts.[2] Sheep, goats, and cattle become affected after they have consumed the plant over several weeks. Many species of agave contain steroidal sapogenins and therefore have the potential for causing poisoning in animals.

Risk Assessment
It is unlikely that household pets would consume enough of the agave leaves to cause liver toxicity. However, the sapogenins may stimulate excessive salvation, and the many sharp spines on the leaf margins and tips of the leaves may cause trauma to the mouths of animals that chew on them.

Clinical Signs
Sheep, goats, and cattle are affected after they have consumed the plant over several weeks. In these species, cholestasis and biliary obstruction can cause icterus and signs of secondary photosensitization. Recovery of affected animals depends upon the chronicity of the poisoning, and since there is no specific treatment, affected animals should be kept out of the sun and fed a more nutritious diet.

References
1. Camp BJ, Bridges CH, Hill DW, Patamali B, Wilson S: Isolation of a steroidal sapogenin from the bile of sheep fed *Agave lecheguilla*. Vet Hum Toxicol 1988, 30: 533-535.
2. Kellerman TS, Coetzer JAW, Naude TW: Plant Poisonings and Mycotoxicoses of Livestock in Southern Africa. Oxford University Press, Cape Town, 1990, pp 32-36.

Aglaonema
Family: Araceae

Common Name
Chinese evergreen

Plant Description
This genus of approximately 20 species is indigenous to China and many parts of tropical Asia. A variety of hybrids have been developed as house and garden plants for their attractive foliage, and ability to survive under low light and poor growing conditions in people's homes. The plants have branching stems with numerous fleshy, variegated, lanceolate leaves up to 12 in. in length (Figures 29 and 30). The flowers are usually small and inconspicuous produced as a fleshy spike enveloped by a spathe, like other members of the Araceae family. The fruits are berries that are frequently brightly colored.

Figure 29 *Aglaonema commitatum*

Toxic Principle and Mechanism of Action
Like other members of the Araceae, Chinese evergreens contain oxalate crystals within the stems and leaves.[1,2] (See Dieffenbachia)

Figure 30 *Aglaonema*

Risk Assessment

Chinese evergreens are commonly grown as house plants, and as such pose a minor risk to household pets that chew on the stems and leaves.

Clinical Signs

Animals that chew on or consume leaves of *Agloenema* species will experience the irritant effects of the oxalates causing excessive salvation and vomiting.

Treatment

Unless salivation and vomiting are excessive, treatment is seldom necessary. The plant should be removed or made inaccessible to the pets that are eating the plant.

References

1. Franceschi VR, Horner HT: Calcium oxalate crystals in plants. Bot Rev 1980, 46: 361-427.
2. Genua JM, Hillson CJ: The occurrence, type, and location of calcium oxalate crystals in the leaves of 14 species of Araceae. Ann Bot 1985, 56: 351-361.

Albizia

Family: Fabaceae

Figure 31 *Albizia julibrissin*

Common Name

Mimosa tree, silk tree, pink siris *(Albizia julibrissin)*
The genus *Albizia* is similar to the genus *Mimosa, Acacia,* and *Leucaena,* and the common names are often interchanged.

Plant Description

A genus of about 150 species of vines, shrubs, or trees native to tropical Asia and Africa, *Albizia* species are deciduous, branching, smooth barked, with attractive, bipinnately compound, fern-like leaves, and showy flowers (Figures 31 and 32). The infloresences are rounded heads of densely clustered white to pink flowers that have 4-5 small petals and long, prominent stamens. Legume seed pods are green, turning brown when mature, and containing numerous ovate, flattened seeds.

Toxic Principle and Mechanism of Action

Several alkaloids that are structural analogs of vitamin B6 (pyridoxine) are present in all parts of the plant and antagonize the effects of pyridoxine.[1,2] Pyridoxine is essential to the formation of neurotransmitters such as GABA, and therefore animals with *Albizia* poisoning exhibit exaggerated responses to stimuli, muscle tremors, abnormal neck posture, abnormal backing and turning behavior, and convulsions.[1] Experimentally sheep developed signs of poisoning in 12-24 hours after they were fed 1-1.5% of their body weight in ground pods.[1] The effects of the toxins can be reversed by injecting pyridoxine.[1,2]

Risk Assessment
Albizia julibrissin is a showy small tree that is popular in gardens, and as such there is potential for animals to be affected by the toxic alkaloids present in the leaves and seeds. Other species of African *Albizia* are well known for their poisoning of livestock.

Figure 32 *Albizia julibrissin*

Clinical Signs
Exaggerated responses to verbal and physical stimuli, muscle tremors, abnormal neck postures and opisthotonus, abnormal backing and turning behavior, and convulsions that last for a few minutes only to recur again after a few minutes may occur in animals that have eaten a large quantity of the plants in a short period.[3] Clinical signs can be reversed by giving parenteral injections of pyridoxine.[2]

References
1. Robinson GH, Burrows GE, Holt EM, Tyrl RJ, Schwab RP: Evaluation of the toxic effects of the mimosa *(Albizia julibrissin)* and identification of the toxicant. *In* Toxic Plants and Other Natural Toxicants, Garland T, Barr AC eds. CAB International, New York. 1998, 453-458.
2. Gummow B, Basstienello SS, Labuschagne L, Erasmus GL: Experimental *Albizia versicolor* poisoning in sheep and its successful treatment with pyridoxine hydrochloride. Onderstepoort J Vet Sci 1992, 59: 111-118.
3. Burrows GE, Tyrl RJ: Toxic Plants of North America. Iowa State University Press, Ames. 2001, 508-510.

Aleurites

Family: Euphorbiaceae

Figure 33 *Aleurites* leaves

Common Name
Tung nut, tung oil tree, wood oil tree, mu oil tree, lumbang nut, China wood oil tree, Japan oil tree, orchid tree, country walnut, Indian walnut, Otaheite walnut

Plant Description
Generally *Aleurites* species are fast-growing, medium-sized trees with a milky sap that are indigenous to the tropical areas of Southeast Asia, Indonesia, Australia, and the Pacific Islands. One species, *A. moluccana* (kukui), is the Hawaiian state tree. In the continental United States, trees are found commonly in Florida and along the Gulf states. Comprising six species, most have large, simple, alternate, long petioled, often lobed leaves that are deciduous (Figure 33) Flowers are produced in panicles, terminally on branches, each flower containing 2-5 sepals and 5-7 petals, that are white, or with red or green veins. Flowers appear before the leaves in the spring. The fruits are ovoid drupes containing 2-5 seeds. The seeds are hard and contain a white flesh.

Toxic Principle and Mechanism of Action
As with other members of the euphorbia family, the *Aleurites* species contain tigliane diterpenoid esters that are irritants.[1,2] The leaves, fruits, and nuts are all poisonous with *A. fordii*, being the most toxic of the species.[1,2] Poisoning is seldom fatal.

Risk Assessment
As the Tung nut tree becomes more widespread as a fast-growing tree in tropical gardens and townships, the potential for animals consuming the leaves or nuts increases. Most poisoning has occurred in people who have eaten the nuts.

Clinical Signs
The irritant diterpenoids in the fruits of the Tung nut tree will induce excessive salivation, vomiting, and abdominal pain.
Activated charcoal orally and the use of cathartics may be helpful in removing plant parts from the digestive system. Intravenous fluid therapy may be necessary in patients with severe fluid loss.

References
1. Emmel MW, Sanders DA, Swanson LE. The toxicity of the foliage *Aleurites fordii* for cattle. J Am Vet Med Assoc 101: 136, 1943.
2. Lin TJ, Hsu CI, Lee KH et al. Two outbreaks of acute Tung nut *(Aleurites fordii)* poisoning. J Toxicol, 34: 87-92, 1996.

Allamanda

Family: Apocynaceae

Common Name
Yellow allamanda, allamanda, golden trumpet vine
Purple allamanda *(A. blanchetii)*

Plant Description
Allamanda is comprised of at least 12 shrub or by vine-like species found mostly in the warm temperate and tropical areas of the Americas. Several cultivars of the species have been developed. *Allamanda cathartica* is commonly grown for its large yellow trumpet-like flowers that are produced year-round (Figures 34 and 35). It is a vine-like shrub growing to 15ft. in height, with a milky sap, and glossy, simple, lanceolate, opposite or whorled leaves. The fruits are globose capsules covered with soft spines, turning brown when ripe. Bush allamanda *(A. schottii)* is a more compact plant with smaller yellow flowers. *Allamanda blanchetii* has purple colored flowers.

Figure 34 *Allamanda cathartica*

Figure 35 *Allamanda cathartica*

Toxic Principle and Mechanism of Action
As the name implies, *Allamanda cathartica*, contains various irritant terpenoids and iridoids that have a purgative effect on the gastrointestinal system.[1] All parts of the plant, but especially the fruits and roots are toxic.

Risk Assessment
Allamanda is a common plant in many warm temperate and tropical gardens, and as such has the potential for causing poisoning in animals that eat the plant.

Clinical Signs
Excessive salivation, vomiting and severe diarrhea can be expected from eating the leaves and roots of allamanda. Symptomatic treatment to prevent dehydration may be necessary in severe cases.

References
1. Coppen JJ, Cobb AL: The occurrence of iridoids in *Plumeria* and *Allamanda*. Phyto Chemistry. 22: 125-128, 1983.

Family: Liliaceae

Figure 36 *Allium giganteum* 'Globemaster'

Figure 37 *Allium schoenoprasum*

Common Name
Onion *(Allium cepa)*, chives *(A. schoenoprasum)*, garlic *(A. sativum)*, leeks *(A. porrum)*, shallots, giant allium

Plant Description
At least 700 species of *Allium* exist in the temperate regions of the world. These bulbous biennial or perennial plants are widely cultivated for their nutritional value and some species as colorful garden ornamentals (Figures 36 and 37). Growing from various sized bulbs, that have papery or fibrous outer layers, the leaves are produced basally, and can be long, linear, lanceolate, flat or grooved, terete, hollow or solid. Flowers are produced as terminal umbels, on stalks that are hollow or solid and can range in height from a few inches to 3-4 feet. Flower colors vary from white to yellow, to pink or mauve. Members of the *Allium* genus have a characteristic onion or garlic odor when the leaves or bulbs are crushed.

Toxic Principle and Mechanism of Action
Allium species contain a variety of sulfur containing compounds (alkylcysteine sulfoxides) that are converted to a variety of sulfides, disulfides, trisulfites and thiosulfonates through the action of plant enzymes once the plant tissues are damaged.[1] The typical onion odor is attributed to the disulfides and try sulfides, while the compound that causes lacrimation when peeling onions, is thiopropanol-S-oxide.[2] One compound, N-propyl disulfide, is a highly reactive oxidant that is responsible for oxidizing hemoglobin. This and other similar sulfide compounds compounds act to deplete critical enzymes such as glucose-6-phosphate and the G6P dehydrogenase that are critical in the cell membrane integrity.[2,3] Once the hemoglobin is oxidized, Heinz bodies form in the red cells, and the defective erythrocytes are removed by the spleen and reticulo-endothelial system. The resulting anemia causes generalized weakness, and can become severe enough to cause fatalities.

Poisoning from onions may occur if animals are fed whole raw onions, chopped, dehydrated, and cooked onions, or products containing onion powder. The toxicity of onions varies depending on the type of onion, growing conditions, the total amount consumed, and the animal species involved. Cats are particularly susceptible to onion poisoning, having been poisoned after being fed baby food containing as little as 0.3% onion powder.[4,5] Dogs are more tolerent of onions. An acute hemolytic anemia was reported in a dog that ingested 3-4 oz. of dehydrated onions.[6] Dogs fed 5.5 g per kg body weight of minced dehydrated onions, exhibited severe hematologic changes within 24 hours exposure.[7] In another study, toxic dose of raw onions in dogs has been cited as equal to or greater than 0.5% of the animal's body weight.[8] Cattle are particularly susceptible to onion poisoning, if they had diet contains greater than 20% onions.[9,10] Horses are less susceptible to onion poisoning,[11] and sheep appear to be able to adapt to rations comprising 100% onions.[12]

Risk Assessment

Most onion poisoning of dogs and cats results from the feeding of raw or cooked onions, or other human foods containing onion products. The risk therefore of onion poisoning in household pets is significant, especially if they are fed table scraps. Cats, in particular should not be fed any food products containing onions. Certain breeds of dog, such as the Akita, have a heritable predisposition to the hemolytic effects of onions.[13] Care should be taken to ensure that onion bulbs, whether they are of the edible or ornamental species, are stored to prevent access by household pets. Similarly, cull or spoiled onions should not be thrown on the compost pile, where dogs have access to them.

Clinical Signs

The onset of clinical signs, following the ingestion of the toxic dose of onions, may vary from one to several days, depending on the total dose of onion consumed. Infected animals become weak . the tactic anorexic and recumbent due to the developing anemia. Heart rate and respiratory rate are increased, and the mucous membranes are pale and can be jaundiced. Frequently the animal's breath smells of onion. The urine may be brown or coffee colored, indicating hemoglobinurea. Examination of the blood will reveal a decreased packed cell volume, and the presence of Heinz bodies in the red blood cells. Methemoglobin levels may also be significantly increased.

Treatment

Severely anemic animals may require a blood transfusion. Generally, however, if the animals are taken off the of the onions and are not stressed, the hemoglobinurea will resolve in 1-3 days and the packed cell volume will return to normal in 2-3 weeks. Inducing vomiting is effective if the onion has been consumed within the last 2 hours. Activated charcoal is indicated after vomiting has stopped. Pet foods containing propylene glycol should be avoided as it enhances Heinz body formation.[14]

References
1. Block E et al: *Allium* chemistry: HPLC analysis of thiosulfinates from onion, garlic, wild garlic (Ramsoms), leeks, scallions, shallot, elephant (great-headed) garlic, chive, and Chinese chive. Uniquely high allyl to methyl ratios in some garlic samples. J Agric Food Chem. 40: 2418-2428, 1992.
2. FenwickGR, Hanley AB: The genus *Allium* – part 2. CRC Crit Rev Food Sci Nutr 22: 272-377, 1985.
3. Yamato O, Hayashi M, Yamasaki M, Maede Y: Induction of onion-induced hemolytic anemia in dogs with sodium propylthiosulfate. Vet Rec. 142: 216-219. 1998.
4. Robertson JE, Christopher MM, Rogers QR: Heinz body formation in cats fed baby food containing onion powder. J Am Vet Med Assoc. 212: 11260-1266, 1998.
5. Kobayashi K: Onion poisoning in the cat. Feline Pract. 11: 22-27, 1981.
6. Farkas MC, Farkas JN: Hemolytic anemia due to ingestion of onions in a dog. J Am An Hosp Assoc. 156: 328-330, 1970
7. Harvey JW, Rackear D: experimental onion-induced hemolytic anemia in dogs. Vet Pathol 22: 387-392, 1981.
8. Kingsbury JM: Poisonous plants of the United States and Canada. Prentice-Hall, Englewood Cliffs, N. J., 1964.
9. Lincoln SD, Howell ME, Combs JJ, Hinman DD: Hematologic effects and feeding performance in cattle fed cull domestic onions *(Allium cepa).*J Am Vet Med Assoc. 200:1090-1094, 1992.
10. Van Der Kolk JH: Onion poisoning in a herd of dairy cattle. Vet Rec. 147: 517-519, 2000.
11. Pierce KR, Joyce JR, England RB, Jones LP: Acute hemolytic anemia caused by wild onion poisoning in horses. J AM Vet Med Assoc 160: 323-327, 1972.
12. Knight AP, et al: Adaptation of pregnant ewes to an exclusive onion diet. Vet Human Toxicology 42: 1-4, 2000.
13. Yamato O, Maede Y: Susceptibility to onion-induced hemolysis in dogs with hereditary high erythrocyte reduced glutathione and potassium concentrations. Am J Vet Res 53: 134-137, 1992.
14. Christopher MM, Perman V, Eaton JW: Contribution of propylene glycol-induced Heinz body formation to anemia in cats. J AM Vet Med Assoc 194: 1045-1056, 1989.

Family: Araceae

Common Name

Elephant's ear, giant elephant's ear, giant taro

Plant Description

Alocasia is a genus of some 70 species originating in tropical Southeast Asia. These perennial large-leafed rhizomatous and tuberous perennials have arrow or heart-shaped leaves from 8-36 in. in length depending on the species. That leaves are long stemmed with distinct red all purple markings. The arum-like flowers are produced on long stems that are frequently obscured by the foliage. The fruits consist of red or orange berries The roots of a few species of alocasia are edible, but the majority are poisonous. *Alocasia* are closely related to taro (*Colocasia* spp.), the tuberous roots of which are commonly used for human food in tropical areas. *Alocasia* species are frequently grown as foliage plants in the garden and as a potted plant in the household (Figures 38 and 39).

Figure 38 *Alocasia macrorrhiza*

Toxic Principle and Mechanism of Action

Like other members of the Araceae family, alocasias contain oxalate crystals in the stems and leaves.[1] The calcium oxalate crystals (raphides) are contained in specialized cells referred to as idioblasts.[2,3] Raphides are long needle-like crystals bunched together in these

Figure 39 *Alocasia sanderiana*

specialized cells, and when the plant tissue is chewed by an animal, the crystals are extruded into the mouth and mucous membranes of the unfortunate animal. The raphides once embedded in the mucous membranes of the mouth cause an intense irritation and inflammation. Evidence exists to suggest that the oxalate crystals act as a means for introducing other toxic compounds from the plant such as prostaglandins, histamine, and proteolytic enzymes that mediate the inflammatory response.[4] (See *Dieffenbachia* species)

Risk Assessment
Alocasia species are frequently grown for their striking foliage and consequently, household pets have access to the plants.

Clinical Signs
Dogs and cats that chew repeatedly on the leaves and stems of alocasia may salivate excessively and vomit as a result of the irritant effects of the calcium oxalate crystals embedded in their oral mucous membranes. The painful swelling in the mouth may prevent the animal from eating for several days. Severe conjunctivitis may result if plant juices are rubbed in the eye.

Treatment
Unless salvation and vomiting are excessive, treatment is seldom necessary. Anti inflammatory therapy may be necessary in cases where stomatitis is severe. The plants should be removed or made inaccessible to the animals that are eating them.

References
1. Lin T-J et al: Calcium oxalate is the main toxic component clinical presentations of *Alocasia macrorrhiza* (L) Schott & Endl poisonings. Vet Hum Toxicol 14: 93-95, 1998
2. Franceschi VR, Horner HT: Calcium oxalate crystals in plants. Bot Rev 46: 361-427, 1980.
3. Genua JM, Hillson CJ: The occurrence, type, and location of calcium oxalate crystals in the leaves of 14 species of Araceae. Ann Bot 56: 351-361, 1985.
4. Saha BP, Hussain M: A study of the irritating principle of aroids. Indian J Agric Sci 53: 833-836, 1983.

Aloe

Family: Aloaceae

Figure 40 *Aloe verra*

Common Name
Aloe, aloe vera, Barbados aloe, candelabra plant, torch plant, uguentine cactus

Plant Description
Aloes are indigenous to Africa, Madagascar and parts of Arabia. This large family of 7 genera, and some 700 species, has considerable variability in the shape and size of the species. Often mistakenly identified as cacti, the aloes are evergreen plants with distinctive rosettes of sword shaped leaves, often with sharp teeth/spines along the leaf blade margins. Some species have thick succulent variegated leaves in dense rosettes, while the tree-like aloes have

rosettes of leaves terminally on the branches. The inflorescence can be a single or branched spike produced from the leaf axil. Individual flowers are tubular or narrowly bell shaped, and vary in color from pale white-green to orange-red. Oval fleshy fruits form after the flowers and turn brown when ripe. Aloes hybridize freely and consequently many ornamental hybrids have been developed for use in the garden and home (Figures 40 and 41).

The most commonly grown species is *Aloe verra (A.barbadensis)*. The leaves contain a thick syrupy juice that has found wide use in various cosmetics and shampoos.[1]

Figure 41 *Aloe secundiflora*

Toxic Principle and Mechanism of Action

Only the genus *Aloe* is known to be toxic, and only a few species in the genus have been studied for their toxicity. Aloes contain varying concentrations of anthraquinone glycosides, the most important of which are barbaloin and homonatoloin.[2] These bitter tasting compounds that are concentrated in the latex of the new leaves are potent purgatives. The anthracene glycosides are not particularly toxic, but are metabolized by intestinal bacteria into more potent compounds such as aloe-emodin.[3] *Aloe candelabrum* and *Aloe ferox* have found commercial use as a purgative known as bitter aloes, or cape aloes. The purgative effects of aloes has been attributed to the production of prostaglandins, and increased activity of colonic mucosal adenyl cyclase. This increases mucus secretion and water content of the colon, which stimulates peristalsis and a resulting diarrhea.[4] In addition to their purgative effects, aloes also have carcinogenic and abortifacient properties.[5,6] The compounds responsible for the supposed beneficial effects of aloe preparations for treating burns and other superficial skin diseases have not been determined. There is evidence that *Aloe vera* gel when applied topically to second degree burns in guinea pigs actually delayed healing of the burns.[7] The cytotoxic effects of low molecular weight compounds in *Aloe vera* warrant caution in its indiscriminant use.[8]

Risk Assessment

Aloes are commonly grown as potted houseplants or garden plants in dry tropical environments. Consequently, household pets have ready access to the plants. Until proven otherwise, all aloes should be considered potentially toxic. Aloe vera products made from the leaf pulp or gel are of low toxicity since the toxic fractions are in the latex of the leaves.

Clinical Signs

Animals that chew on the fleshy leaves of aloes can ingest sufficient barbaloin that can result in severe purgation. Unless animals become severely dehydrated as a result of the diarrhea, treatment is seldom necessary. If the animal's urine is alkaline, the barbaloin will cause the urine to turn red.

Urticaria and dermatitis has been reported in some people who have applied topically or taken orally *Aloe vera* products.[9]

References

1. Grindlay DL, Reynolds T: The *Aloe* vera phenomenon: a review of the properties and modern uses of the leaf parenchyma gel. J Ethnopharmacol 16: 117-151, 1986.
2. Groom QJ, Reynolds T: Barbaloin in aloe species. Planta Med 53: 345-348, 1987.
3. Robinson T: the organic constituents of high and plants. Cordus Press, North Amherst, Mass, 1991.
4. Capasso F, Mascolo N, Aulore G, Duraccio MR: Effect of indomethacin on aloin and 1,8-dihydroxyantharaquinone-induced production of prostaglandins in rat isolated colon. Prostaglandins 26: 557-562, 1983.
5. Watt JM, Breyer-Brandwijk MG: The medicinal and poisonous plants southern and eastern Africa, 2nd Ed, E & S Livingston, Edinburg, pp 679-687, 1962.
6. Wolfe D, Schmutte C, Westendorf J, Marquardt H: Hydroxyanthraquinones as tumor promoters: enhancement of malignant transformation of mouse fibroblasts and growth stimulation of primary rat hepatocytes. Cancer Res 50: 6540-6544, 1990.
7. KaufmanT, Kalderon N, Ullman Y, Berger J: Aloe vera gel hinted wound healing of experimental second degree burns: a quantitative controlled study. J Burn Care Rehab 9: 156-162, 1988.
8. Avila H, Rivero J, Herrera F, Fraile G: Cytotoxicity of a low molecular weight fraction from *Aloe vera* (*Aloe barbadensis* Miller) gel. Toxicon 35:1423-1430, 1997.
9. Hogan DJ: Widespread dermatitis of the topical treatment of chronic leg ulcers and stasis dermatitis. Can Med Assoc J 138: 336-339, 1998.

Amaryllis belladonna

(Synonym: Callicore rosea)
Family: Liliaceae (Amaryllidaceae)

Common Name
Belladonna lily, naked lady,
Resurrection lily

Plant Description
Consisting of a single species indigenous
to Southwestern Africa, *Amaryllis
belladonna* is a lily growing from a large
bulb. White to rosy pink trumpet-shaped,
scented, showy flowers are produced in the
fall on tall stems up to 30 in. in length
before the basal strap like leaves emerge
(Figure 42). The succulent leaves appear
after the flowers wither. Hybrids of
Amarylis and Crinum (*Amacrinum
memoria-corsii*) are very similar to the
species (Figure 43). Various cultivars exist.

Figure 42 *Amaryllis belladonna*

Toxic Principle and Mechanism of Action
Several phenanthridine alkaloids
including lycorine, belarmine, ambelline,
and undulatine have been identified in
the leaves, stems and bulbs of *Amaryllis
belladonna.*[1] The phenanthridine alkaloids
are present in many of the Liliaceae, most
notably in the *Narcissus* group (see
Narcissus). The alkaloids have emetic,
hypotensive and respiratory depressant
effects, and cause excessive salivation,
abdominal pain, and diarrhea. Calcium
oxalate raphides may also contribute to
digestive symptoms.

Figure 43 *Amacrinum memoria-corsii*

As many as 15 other phenanthridine alkaloids have been isolated from other
genera of the Amaryllis family, including species of *Clivia, Galanthus,
Haemanthus, Hippeastrum, Hymenocallis, Leucojum, Narcissus, Nerine, Sprekelia,*
and *Zephranthes*

Risk Assessment
This attractive garden plant, and occasionally potted houseplant has rarely caused
poisoning in animals, but it has the potential to do so.

Clinical Signs

Vomiting, excessive salivation, abdominal pain, diarrhea, and difficulty in breathing, are associated with the phenanthridine alkaloids present in the lily family. If large quantities of the leaves and bulb are consumed, depression, ataxia, seizures, and hypotension may develop. Poisoning is rarely fatal, and can generally be treated symptomatically.

References

1. Martin SF: The Amaryllidaceae alkaloids. In The Alkaloids: Chemistry and Physiology. vol 30, Brossi A (ed) Academic Press, San Diego, Calif 251-376, 1987.

Anemone

Family: Ranunculaceae

Figure 44 Anemone coronaria

Figure 45 Anemone sylvestris

Common Name

Anemone, windflower, wind poppy

Plant Description

There are approximately 150 species of *Anemone* occurring widely in the northern hemisphere and especially in temperate Asia. There are approximately 25 to 30 native and introduced species of *Anemone* in North America. These perennial plants have palmate, basal leaves that are divided into a few or many leaflets. Plants arising from rhizomes or tubers can be generally grouped into spring flowering and fall flowering varieties. Flowers have five or more petals and come in a wide spectrum of colors. The spring flowering varieties of Anemone are typified by *Anemone coronaria*. (Figure 44). The Japanese anemone, and its hybrids are spreading perennials with fibrous roots that bloom in the fall (Figure 45).

Toxic Principle and Mechanism of Action

Anemone species contain the irritant glycoside ranunculin that is converted to protoanemonin when the plant tissues are chewed and macerated.[1] Protoanemonin levels amongst the species vary, and appear to be in the range of 0.02-0.05% of the green plant.[2] Protoanemonin is the vesicant, and it is polymerized to the non toxic anemonin. The dried plant contains mostly anemonin and is therefore not toxic.

Risk Assessment

Anemones are not a significant problem to household pets as the bitter irritant effects of the plants are a deterrent to most dogs and cats. However, the showy Anemone's that are sold as potted plants or as garden ornamentals have the potential to be chewed and eaten by pets.

Clinical Signs

Excessive salivation, vomiting and diarrhea can be anticipated if anemones are eaten. Treatment if necessary would be symptomatic (See Ranunculus).

References

1. Hill R, Van Heyningen R: Ranunculin: the precursor of the vesicant substance I Buttercup. Biochem J 49: 332-335, 1951.
2. Bonora A, Dall'Olio G, Bruni A: Separation and quantification of protoanemonins in Ranunculaceae by normal and reversed phase HPLC. Planta Med 51: 364-367, 1985.

Anthurium

Family: Araceae

Common Name

Flamingo flower, painter's palette, pigtail plant, Lenguna de Vaca, anturio

Plant Description

Native to tropical America, *Anthuriums* belong to a large and diverse genus of tropical plants in the arum family. The species that are cultivated and used most commonly by florists have bright red flat spathes carried above the glossy green, leathery leaves. The flowers are actually very small and clustered on the spadix which is surrounded by the flat spathe (Figures 46 and 47). Hybridization of the species has resulted in various cultivars with spathes that are white, yellow, or mottled.

Figure 46 *Anthurium andraeanum*

Toxic Principle and Mechanism of Action

Like other members of the arum family the anthuriums contain calcium oxalate crystals in the leaves and stems that can cause inflammationa and pain in the mouth of animals or people who chew on the plant parts.[1,2] (see *Dieffenbachia* species)

Risk Assessment

As they are common house and garden plants, anthuriums have the potential for causing poisoning in household pets that might chew upon the plant.

Figure 47 *Anthurium spathe* and *spadix*

Clinical Signs

Dogs and cats that chew repeatedly on the leaves and stems of anthuriums may salivate excessively and vomit as a result of the irritant effects of the calcium oxalate crystals embedded in their oral mucous membranes. The painful swelling in the mouth may prevent the animal from eating for several days. Severe conjunctivitis may result, if plant juices are rubbed in the eye.

Treatment

Unless salvation and vomiting are excessive, treatment is seldom necessary. Anti inflammatory therapy may be necessary in cases where stomatitis is severe. The plant should be removed or made inaccessible to the animals that are eating the plant

References

1. Genua JM, Hillson CJ: The occurrence, type, and location of calcium oxalate crystals in the leaves of 14 species of Araceae. Ann Bot 56: 351-361, 1985.
2. Spoerke DG, Smolinske SC. Toxicity of House Plants. CRC Press. Boca Raton, Florida, pp 76, 1990

Arisaema

Family: Araceae

Figure 48 *Arisaema triphyllum*

Common Name

Jack in the pulpit, Indian turnip, green dragon, dragon root

Plant Description

There are approximately 150 species of this genus found primarily in temperate tropical parts of the Northern Hemisphere. There are two species found in North America: *Arisaema triphyllum*, and *A. dracontium*. Arising from tuberous or rhizomatous roots, these perennials have palmate leaves with three or more leaflets. The leaves are generally bourn on long stems emerging from the base of the plant. Inflorescences are typical of the Arum family, being solitary, and arising on long stems (Figures 48 and 49). The central spathe is made up of numerous small fleshy flowers, surrounded by a greenish white spathe. Considerable variation exists in the size, color, and shape of the spathe depending on the species. The flowers are followed by fleshy red fruits. This plant is commonly grown as ornamental for its foliage and interesting flowers.

Figure 49 *Arisaema* fruits

Toxic Principle and Mechanism of Action

Like other members of the Araceae family, *Arisaema* contain oxalate crystals in the stems and leaves. The calcium oxalate crystals (raphides) are contained in specialized cells referred to as idioblasts.[1,2] Raphides are long needle-like crystals grouped together in these specialized cells, and when the plant tissue is chewed by an animal, the crystals are extruded into the mouth and mucous membranes of the unfortunate animal. The raphides once embedded in the mucous membranes of the mouth cause an intense irritation and inflammation. Evidence exists to suggest that the oxalate crystals act as a means for introducing other toxic compounds from the plant such as prostaglandins, histamine, and proteolytic enzymes that mediate the inflammatory response.[3]

Risk Assessment

Arisaema species are frequently grown for their striking foliage and unique flowers, and consequently, household pets may be poisoned if they chew or swallow the plants.

Clinical Signs

Dogs and cats that chew repeatedly on the leaves and stems of *Arisaema* salivate excessively and vomit as a result of the irritant effects of the calcium oxalate crystals embedded in their oral mucous membranes. The painful swelling in the mouth may prevent the animal from eating for several days. Severe conjunctivitis may result, if plant juices are rubbed in the eye.

Treatment

Unless salvation and vomiting are excessive, treatment is seldom necessary. Anti inflammatory therapy may be necessary in cases were stomatitis is severe. The plant should be removed or made inaccessible to the animals that are eating the plant.

References

1. Franceschi VR, Horner HT: Calcium oxalate crystals in plants. Bot Rev 46: 361-427, 1980.
2. Genua JM, Hillson CJ: The occurrence, type, and location of calcium oxalate crystals in the leaves of 14 species of Araceae. Ann Bot 56: 351-361, 1985.
3. Saha BP, Hussain M: A study of the irritating principle of aroids. Indian J Agric Sci 53: 833-836, 1983.

Family: Brassicaceae

Figure 50 *Armoracia rusticana*

Figure 51 *Armoracia rusticana* root

Common Name
Horseradish, red cole

Plant Description
Native to south eastern Europe, Armoracia rusticana has become widely cultivated for the taproot which is used to make horseradish sauce. Two other species of Armoracia grow in Europe and Siberia but are not common.

A vigorous herb with large 12-18 inch light to dark green leaves with a puckered surface. Loose panicles of 4 petalled, white flowers are produced in summer (Figure 50). The plant is a prolific seed producer, and becomes invasive. The white taproot is harvested to make horseradish sauce (Figure 51).

Japanese horseradish or wasabi is not produced from *Armoracia* species, but rather from the separate genus *Wasabia*.

Toxic Principle and Mechanism of Action
Armoracia species contain glucosinolates, the best know of which are sinigrin and 2-phenylethyl glucosinolates. The root and the seeds contain the highest concentrations. The glucosinolates are rapidly hydrolysed to ally-isothiocyanate which is a strong irritant.[1] In low concentrations glucosinolates are appetite stimulants, but in high concentration they are potent irritants especially if they get into the eyes.

Risk Assessment
Horseradish although commonly grown in vegetable gardens is not of great risk to household pets. However, the root once harvested and brought into the kitchen it can become a hazard to dogs that might chew and eat it.

Clinical Signs
Reports of poisoning in animals from eating horseradish are limited to livestock where apparently the horseradish caused gastric inflammation, colic, and death.[2] Mouth, upper respiratory distress, and gastric irritation are commonly reported in

humans unaccustomed to eating horseradish. In severe cases some individuals develop temporary "horseradish syncope" and collapse from vasomotor collapse.[2]

References
1. Fenwick GR, Heaney RK, Mawson R: Glucosinolates. In Toxicants of Plant Origin, vol2, Glycosides. Cheeke PR ed, CRC Press, Boca Raton, Florida. pp 1-41, 1989.
2. Spoerke DG, Smolinske SC. Toxicity of House Plants. CRC Press. Boca Raton, Florida. pp 77-78, 1990.

Arum

Family: Araceae

Common Name
Arum, cuckoo-pint, Lords and Ladies,

Plant Description
There are approximately 25 species of Arum indigenous to the Mediterranean area and as far north as England. Some species are cultivated in North America. Tuberous perennials with broad fleshy, arrow head-shaped, often variegated, leaves, and long petioles. The inflorescence consists of a spathe, the true minute flowers are densely arranged on the central spadix, that is surrounded by a large white, green-yellow, or purple spathe. The fruits are orange or red berries (Figure 52).

Figure 52 *Arum maculatum* fruits

The white "arum lily" commonly sold in flower shops is not a member of the Arum genus, but is *Zantedeschia aethiopica*.

Toxic Principle and Mechanism of Action
Like other members of the Araceae family, *Arum* species contain oxalate crystals in the stems and leaves. The calcium oxalate crystals (raphides) are contained in specialized cells referred to as idioblasts.[1,2] Raphides are long needle-like crystals grouped together in these specialized cells, and when the plant tissue is chewed by an animal, the crystals become embedded in the mouth and mucous membranes of the unfortunate animal. The raphides once embedded in the mucous membranes cause an intense irritation and inflammation.

Risk Assessment
Arum species are frequently grown for their striking foliage and unique flowers, and consequently, household pets may be poisoned if they chew or swallow the plants.

Clinical Signs

Dogs and cats that chew repeatedly on the leaves and stems of *Arum* salivate excessively and vomit as a result of the irritant effects of the calcium oxalate crystals embedded in their oral mucous membranes. The painful swelling in the mouth may prevent the animal from eating for several days. Severe conjunctivitis may result, if plant juices are rubbed in the eye.

Treatment

Unless salvation and vomiting are excessive, treatment is seldom necessary. Anti inflammatory therapy may be necessary in cases where stomatitis is severe. The plant should be removed or made inaccessible to the animals that are eating the plant.

References

1. Franceschi VR, Horner HT: Calcium oxalate crystals in plants. Bot Rev 46: 361-427, 1980.
2. Genua JM, Hillson CJ: The occurrence, type, and location of calcium oxalate crystals in the leaves of 14 species of Araceae. Ann Bot 56: 351-361, 1985.

Asclepias

Family: Asclepiadaceae

Figure 53 *Asclepias tuberosa*

Common Name
Milkweed, butterfly weed, blood flower.

Plant Description
Existing mostly in the Americas, milkweeds are made up of about 150 species. The name *Asclepias* is derived from Asklepios, the Greek god of healing. Most are wild flowers growing in a variety of habitats, and a few have been cultivated as ornamentals because of their showy flowers, eg: Butterfly weed *(Asclepias tuberosa),* swamp milk weed *(A. incarnata)* and blood flower *(A. curassavica)* (Figures 53-55).

Asclepias species are herbs, generally with a white milky sap, erect or decumbent, branced or unbranched stems, and leaves that are alternate or whorled, narrow (verticillate) or elliptical to lanceolate. Inflorescences are terminal or arising from leaf axils. Flowers have fused sepals and petals and are spreading or reflexed. Colors vary from white, greenish-white, pink, orange, and red depending upon the species. Fruits are fusiform to globose or ovoid follicles (Figures 56 and 57). The many flat seeds are attached to silky white hairs that aid in wind dispersion.

Figure 54 *Asclepias incarnata*

Toxic Principle and Mechanism of Action

Asclepias species can generally be considered to be neurotoxic or cardiotoxic, although a few species have both toxic properties. The cardiotoxic species are generally those with leaf blades greater than 3.5cm. in width, while those with narrow, grass-like leaves tend to be neurotoxic.[1] The cardiotoxic broad leafed species contain cardenolides which inhibit Na$^+$ and K$^+$ ATPase, critical in normal myocardial function. Some species that contain high levels of cardenolides include *Asclepias asperula, A. labriformis, A. eriocarpa,* and *A. curassavica.*[2-4] The action of the cardenolides is similar to that of ouabain and digitalis and may induce cardiac conduction disturbances, arrhythmias, and heart block at toxic doses.[3] The monarch butterfly larvae are well known for their preference for the more toxic broad leafed milkweeds, and have been shown to accumulate the cardenolides in their skin thereby protecting them and the resulting monarch butterfly from predation by birds.[5] Similar immunoreactive cardiac glycosides detectable by radioimmunoassay using antibodies to cardiac glycosides are detectable in other plant genera including *Nerium, Thevetia, Ackocanthera, Calotropis,* and *Cryptostegia* species.[6]

The verticillate or narrow leafed species such as the whorled milkweed *(A. subverticillata),* the eastern whorled milkweed *(A. verticillata)* and the plains milkweed *(A. pumilla)* are neurotoxic with little or no cardenolide content.[1,2] The toxin(s) responsible for the neurologic signs has not been defined. The neurotoxin(s) appear to be cumulative in effect and induce severe colic, muscle tremors, incoordination, seizures and respiratory failure prior to death.

Figure 55 *Asclepias curassavica*

Figure 56 *Asclepias tuberosa* flowers and pods

Figure 57 *Asclepias fruiticosa*

Risk Assessment

Milkweeds such as the butterfly weed *(A. tuberosa)*, and swamp milkweed *(A. incarnata)*, are commonly grown as garden plants for their showy flowers. The blood flower milkweed *(A. curavassica)*, a South American species has become a common garden plant in more tropical areas. Even the indigenous milk weeds such as the showy milkweed *(Asclepias speciosa)*, and the desert milkweed *(A. subulata)* are popular in wildflower gardens as these are favorites of the monarch butterfly caterpillars that feed on the milkweeds exclusively. In general the milkweeds pose little risk to household pets as the milky sap makes the plants distasteful. Most poisoning from milkweeds occurs in cattle, sheep and horses that eat the narrow leafed species of milkweed such as the whorled milkweed *(A. subverticillata)*. The narrow leafed species of milkweed remain toxic even when dried in hay. The broad leafed species tend to have leathery leaves that are unpalatable to most animals. Some African species of milk weed, such as *A. physocarp*, and *A. fruiticosa* (balloon cotton) (Figure 57), which are occasionally grown as ornamentals for their unusual bladder like pods, are also poisonous.[7]

Clinical Signs

Milkweeds that contain cardenolides induce signs of digestive upset, including colic, diarrhea, followed by depression, irregular respiration, and death.[2,7] Cardiac irregularities although anticipated with the cardenolides are rarely observed. Seizures are not observed, and death occurs without convulsions. In contrast the neurotoxic milkweeds induce muscle tremors, colic, incoordination, inability to stand, followed by seizures, convulsions, respiratory failure and death.

If recognized early enough in the course of poisoning, activated charcoal orally, atropine, and other antiarrythmic drugs may be used to counter the cardiac toxicity of the cardenolides. Symptomatic treatment should be provided for the animal showing neurotoxicity.

References

1. Ogden L, Burrows GE, Tyrl RJ, Ely RW: Experimental intoxication in sheep by *Asclepias*. In Poisonous Plants: Proceedings of the Third International Symposium. James LF, Keeler RF, Bailey EM, Cheeke PR, Hegarty MP eds, Iowa State University Press, Ames. pp 495-499, 1992.
2. Burrows GE, Tyrl RJ: Toxic Plants of North America. Iowa State University Press, Ames. pp 125-146, 2001.
3. Benson JM, Seiber JN, Keeler RF: Studies on the toxic principle of *Asclepias eriocarpa*, and *Asclepias labriformis*. In Effects of Poisonous Plants on Livestock. Keeler RF, Van Kampen KR, James LF eds. Academic Press NY. pp 273-284, 1978.
4. Seiber JN et al: New cardiac glycosides (cardenolides) from *Asclepias* species. In Plant Toxicology – Proceedings of the Australia-USA Poisonous Plants symposium. Seawright AA, Hegarty MP, James LF, Keeler RF eds. Animal Research Institute, Yeerongpilly, Brisbane, Aust 427-437, 1985.
5. Malcolm SB, Cockrell BJ, Brower LP: Cardenolide fingerprint in monarch butterflies reared on common milkweed, *Asclepias syriaca* L. J Chem Ecol 15: 819-853, 1989.
6. Radford DJ, Gillies AD, Hinds JA, Duffy P: Naturally occurring cardiac glycosides. Med J Austr 144: 540-544, 1986.
7. Watt JM, Breyer-Brandwijk MG: The medicinal and poisonous plants of southern and eastern Africa. E & S Livingston, Edinburgh 5119-141, 1962.

Asparagus

Family: Liliaceae

Common Name
Asparagus, asparagus fern, emerald feather

Plant Description
A large genus of 50-60 species from Europe and Asia, best known for the edible species, *Asparagus officinalis*. There are many species of Asparagus, that are cultivated as house and garden plants because of their fernlike foliage (Figure 58). Asparagus species have a rhizomatous roots with erect, spreading or climbing stems, bearing small scale-like, wedge-shaped leaves or thin branches or stems that function as leaves. The greenish yellow or white flowers are born singly or clusters. Fruits are bright red berries with 1-6 seeds.

Figure 58 *Asparagus densiflora*

Toxic Principle and Mechanism of Action
Various *Asparagus* species have been reported to contain sapogenins and other compounds with irritant, cardiac and sedative effects.[1,2]

Risk Assessment
Although *Asparagus* species are commonly grown in vegetable gardens for their edible spring time shoots, and others are grown as a potted plants for their attractive foliage, poisoning of children, or household pets from eating Asparagus berries is rare. Consumption of Asparagus berries is often listed by poison control centers as one of the more commonly ingested houseplants.[3]

Clinical Signs
Depending on the Asparagus species and the quantity of berries consumed, the anticipated clinical signs might include salivation vomiting and diarrhea. Symptomatic treatment may be necessary in the more severely affected animal.

References
1. Fuller TC, McClintock E: Poisonous Plants of California. University, California Press, Berkeley pp 289-292, 1986.
2. Kar OK, Sen S: contents of sapogenins in the diploid, tetraploid, and hexaploid asparagus. Int J Crude Drug Res 24: 131-133, 1986.
3. Veltri JC, Temple AR: 1983 annual report of the American Association of Poison Control Centers national data collection system. Am J Emerg Med 2: 420-424, 1984.

Family: Solanaceae

Figure 59 *Atropa belladonna*

Common Name
Deadly nightshade, belladonna

Plant Description
A genus of four species from Europe, Asia, and North Africa, *Atropa belladonna* is the best-known and most frequently cultivated. Perennial, much branched plants growing from 3-6 ft. in height, with soft ovate leaves. Flowers are bell shaped, nodding, purplish to dull red in color, produced singly at or near branch tips. In the Fall shiny black berries sitting on a large persistent calyx are produced (Figure 59).

Toxic Principle and Mechanism of Action
All parts of the plant contain quantities of tropane alkaloids, including hyoscine (scopolamine), hyoscyamine, and norhyoscine.[1] Other common plant genera containing similar tropane alkaloids included *Brugmansia, Datura,* and *Hyoscyamus.* The tropane alkaloids antagonize the actions of acetylcholine and muscarinic, cholinergic receptors, and therefore have a profound effect upon the autonomic nervous system involving the heart, the digestive system, the eye and the central nervous system. The tropane alkaloids have hallucinogenic properties that have led to human abuse and fatalities. Deadly nightshade has a long history of human medicinal use and toxicities, and acquired the name belladonna meaning beautiful lady, because extracts from the plant were used to cause dilation of the pupils in women to enhance their beauty.[2]

Risk Assessment
Deadly nightshade is rarely grown as an ornamental in North America and is therefore unlikely to be a significant risk to children or household pets. In Europe, deadly nightshade poisoning continues to be one of the most common severe plant intoxications of people primarily because it is eaten for its hallucinogenic properties.[3] The black fruits are a primary cause of poisoning.

Clinical Signs
The early signs of deadly nightshade poisoning include depression, unusual behavior, hallucinations, weakness, tachycardia, mydriasis, dry mucous membranes and constipation. Respiratory failure develops in severe cases. Physostigmine as a cholinergic drug may be used to counter the effects of the tropane alkaloids.

References
1. Schreiber K, Steroid alkaloids: the subtle and group. In the Alkaloids, vol 10, Manske RHF ed. Academic press, New York pp1-192, 1968
2. Hass LF: *Atropa belladonna* (deadly nightshade). J Neurol Neurosurg Psychiatry 58: 283, 1995.
3. Schneider F et al: Plasma and the urine concentrations of atropine after the ingestion of cooked deadly nightshade berries. Clin Toxicol 34: 113-117, 1996.

Aucuba japonica

Family: Cornaceae (Aucubaceae)

Common Name
Aucuba, Japanese aucuba, spotted laurel, gold dust tree, aoki

Plant Description
A genus of three species from eastern Asia grown for their attractive leaves and tolerance of shady conditions. Shrubs grow to 5 m. in height, with green branching stems, opposite, glossy, evergreen, elliptic to ovate, serrate, and yellow to white spotted leaves. Male and female flowers are born on separate plants (dioecious). The male plants are grown for their showy leaves, while the female (pistillate) plants have attractive leaves and star-shaped purple flowers that produce red oblong fleshy fruits or drupes. Numerous cultivars of aucuba have been developed (Figures 60 and 61).

Figure 60 *Aucuba japonica*
(Photographer: Catherine Trendell)

Toxic Principle and Mechanism of Action
Aucubin, an iridoid glycoside, is found in the leaves of the plant and has laxative and diuretic properties.[1,2] Studies have shown aucubin to inhibit hepatic RNA and protein synthesis and thereby have protective properties against liver toxins.[3,4] The irritant effects of the iridoid glycosides have been reported to cause mild diarrhea and vomiting in pets that eat the leaves and stems.[5]

Figure 61 *Aucuba japonica* leaves

Risk Assessment
Aucuba species have only minor potential for poisoning, but they are commonly grown as garden ornamentals in tropical and mild climates, and as house plants in cold climates.

Clinical Signs
Mild vomiting and diarrhea may result from consumption of the plant parts, and are usually self limiting and rarely require supportive treatment.

References

1. Inouye H: Purgative activities of iridoid glucosides. Planta Medica. 25:285-28, 1974.
2. Trim AR, Hill R: The preparation and properties of aucubin, asperuloside, and some related glycosides. Biochem J. 50: 310-318, 1952.
3. Chang IM, Ryu JC, Park YC, Yun HS, Yang KH: Protective activities of aucubin against carbon tetrachloride-induced liver damage in mice. Drug & Chemical Toxicology 6: 443-453, 1983.
4. Chang LM, Yun HS, Kim YS, Ahn JW: Aucubin: potential antidote for alpha-amanitin poisoning. J Toxicology - Clinical Toxicology. 22: 77-85, 1984.
5. Leroux V: Poisoning of pets by house plants. Point Vet 18: 45-55, 1986.

Baptisia

Family: Fabaceae

Figure 62 *Baptisia australis*

Common Name
Blue indigo, blue false indigo, wild indigo

Plant Description
Baptisia is a genus of some 30 species of perennials, native to the prairie and woodland areas of Eastern and Central North America. Arising from a woody rootstock, the erect, solitary branching stems reach a height of 1 m (3 ft.). Leaves are palmately compound, short petioled with 3 obovate leaflets, and entire margins. The inflorescences are terminal or axillary racemes, few to many flowered, each flower with 5 fused sepals and 5 petals. The pea-like flowers may be white, green, yellow, and blue-violet (Figures 62 and 63). The leguminous pods are oblong and conspicuously beaked.

Several species of *Baptisia* are recognized commonly and have been introduced as garden ornamentals includeding:
Baptisia alba-white wild indigo.
B. australis-blue indigo, blue false indigo.
B. bracteata-plains wild indigo.
B. tinctoria-yellow wild indigo

Toxic Principle and Mechanism of Action
Quinolizidine alkaloids similar to those found in *Laburnum* species (Golden chain tree) are found in all parts of the plant.[1] Cytisine is the primary alkaloid present and its primary effect is upon the nicotinic acetylcholine receptors.[2] Consequently, affected animals will show signs of excessive salivation, anorexia, muscle tremors, abdominal pain, increased heart rate, and depressed respiration.

Risk Assessment

Baptisia species pose minimal risk to household pets, but as they are becoming increasingly popular in gardens, and the dry seed pods are used in floral decorations, the potential for poisoning increases.

Clinical Signs

Depending on the quantity of leaves or seeds consumed, affected animals typically show signs of increased salivation, anorexia, incoordination, muscle tremors, and colic. Vomiting without diarrhea, weakness, and tachycardia can be anticipated in dogs and children.[3]

Figure 63 *Baptisia australis* flowers

References

1. Cranmer MF, Mabry TJ: The lupine alkaloids of the genus *Baptisia (Leguminosae)*. Phytochemistry 5: 1133-1138, 1966.
2. Schmeller T, Sauerwein M, Sp one orer F, Wink M, Muller WE: Binding of quinolizidine alkaloids to nicotinic and muscarinic acetylcholine receptors. J Nat Prod 57: 1316-1319, 1994.
3. Burrows GE, Tyrl RJ: Toxic Plants of North America. Iowa State University Press, Ames. pp 535-537, 2001.

Begonia species

Family: Begoniaceae

Common Name

Begonia
With over 1500 species, *Begonia* are native to most tropical and sub-tropical areas of the world except Australia. Most of the 10,000 cultivars of Begonia are derived from a few species and are loosely grouped into cane-stemmed, shrubby, winter-flowering, and tuberous varieties.[1]

Plant Description

A large and diverse genus of perennial

Figure 64 *Begonia* hybrid

small (5-15cm) to tall (3 m) in height branching plants with variably shaped leaves that are succulent-like and brittle. In some hybrids the leaves have attractive markings. Flowers are quite variable depending upon the species, some such as the tuberous begonias have multiple, large attractive blooms. The male and female flowers are produced on the same plant, with the female flowers having broad, colorful, fleshy flanges on the ovaries which develop into winged fruits

Figure 65 *Tuberous begonia flower*

(Figures 64 and 65). The tuberous begonia group tend to be deciduous in the winter and grow back from the tubers.

Toxic Principle and Mechanism of Action

Begonia species contain oxalate crystals especially in the tubers, and also the stems.[2] These crystals can readily penetrate the mucous membranes of the mouth causing inflammation in a similar manner to other oxalate containing plants such as *Dieffenbachia* species. Some species of *Begonia* have been shown to contain cucurbitacins that are cytotoxic.[3]

Risk Assessment

For the most part *Begonia* plants are rarely associated with any poisoning except for a contact dermatitis in some people who handle the plants frequently.[4] The tubers of the tuberous begonias are a potential risk to household pets that gain access to them.

Clinical Signs

Excessive salivation, mouth irritation, and vomiting may be expected if the plants or tubers are chewed or swallowed. (see *Dieffenbachia spp.*)

References

1. The Plant Book. Page S, Olds M (eds) Mynah. 135-137, 2003.
2. Spoerke DG, Smolinske SC. Toxicity of House Plants. CRC Press. Boca Raton, Florida. 82-83, 1990.
3. Doskotch RW. Malik MY. Beal JL. Cucurbitacin B, the cytotoxic principle of Begonia *tuberhybrida* var. alba. Lloydia. 32: 115-122, 1969.
4. Paulsen E. Occupational dermatitis in Danish gardeners and greenhouse workers (II). Etiological factors. Contact Dermatitis 38: 14-19, 1998.

Blighia

Family: Sapindaceae

Common Name
Akee

Plant Description
Consisting of several species of large trees originating from West Africa, one species *Blighia sapida*, has been imported into some of the Caribbean islands and Florida. This large tropical tree has pinnately compound leaves with 6-10 ovate leaflets, with entire margins. The flowers are fragrant, small, white and produced in drooping pannicles. The fruits are three lobed, bell-shaped, 3-4 in. capsules with a thick, leathery rind becoming yellow to red when ripe. The ripe capsules split open to expose 3 glossy black seeds, each attached to a fleshy white to yellow aril (Figure 66).

Figure 66 *Blighia sapida* fruits
(Photographer: Brinsley Burbidge)

Toxic Principle and Mechanism of Action
The highest concentrations of the water soluble, heat stabile, toxic amino acids Hypoglycin A and B are found in the seeds of the unripe fruits.[1] The ripe aril of the fruit has very low if any of the toxins, and is commonly consumed in Jamaica.[2] The hypoglycin A, once metabolized, inhibits many metabolic enzymes in cell mitochondria, that results in severe hypoglycemia.[3]

Risk Assessment
As these trees become more popular ornamentals, the chances of dogs having access to the unripe seeds increases. There are no reports of poisoning in animals to date, while human cases are not unusual.[2]

Clinical Signs
Although most cases of akee poisoningis reported in humans who eat the unripe seeds, similar effects have been produced in cats experimentally. Following ingestion of the unripe seeds, vomiting or drowsiness, followed by seizures and coma are to be expected. Hypoglycemia is marked. Fatal encephalopathy has occurred in children eating the unripe seeds and fruits.[4] Treatment is directed at restoring blood glucose levels, and keeping the patient well hydrated.

References
1. Tanaka K, Kean EA, Johnson B: Jamaican of vomiting sickness: biochemicalinvestigation of two cases. New Eng J Med 295: 461-467, 1976.
2. Morton JF: Plants Poisonous to People in Florida and other warm areas. Southeastern Printing Co. pp 63-64, 1962.
3. Kean EA: Hypoglycin. *In* Toxicants of Plant Origin, vol 3, Proteins and Amino Acids. Cheeke PR ed, CRC press, Boca Raton, Florida pp 229-262, 1989.
4. Meda HA et al: Epidemic of fatal encephalopathy in preschool children in Burkina Faso and consumption of unripe ackee *(Blighia sapida)* fruit. Lancet 353: 536-540, 1999.

Bowiea

Family: Liliaceae

Figure 67 *Bowiea volubilis*

Common Name
Sea onion, climbing onion

Plant Description
A leafless, climbing-stemmed plant of southern and eastern Africa, consisting of 2-3 species. The stems arise from a large, silvery-green bulb, with large scales and papery tunics. The bulb sits mostly above ground, preferring hot, dry conditions. The twining, branched, photosynthetic stems may reach 6-10 feet in length. The small white-yellow, star-like flowers are produced at the ends of the stems (Figure 67).

Toxic Principle and Mechanism of Action
All parts of the plant including the bulb, stems, and flowers contain potent cardiotoxic bufadienolides including hellebriginin, and boviside A. Two grams of the dried flowers supposedly have the same toxicity as 1gm of digitoxin.[1] As with other bufadienolides, cardiac arrhythmias and heart block occur as the toxic dose increases. The bulb also contains raphides of calcium oxalate crystals that can cause skin and mouth irritation if handled or chewed.

Risk Assessment
Bowiea species are rarely grown except by plant hobbyists who value the xeritypic attributes of the plant. It can be grown as a potted house plant and is therefore a risk to animals and children who might chew on it and consume the stems and bulbs. Experimentally *Bowiea* species have been shown to be toxic to dogs, cats, sheep, and goats.[2] Human poisoning has occurred when the plant has been misused in herbal remedies.[1]

Clinical Signs
Severe salivation and vomiting may occur from the calcium oxalate raphides in the bulbs. Decreased heart rate, cardiac arrhythmias, and heart block may result when larger quantities of the plant have been consumed.

Treatment if necessary, is usually achieved by the oral administration of activated charcoal, intravenous fluids, and atropine if cardiac arrhythmias are present.

References
1. Watt JM, Breyer-Brandwijk MG: Liliaceae. *In* The Medicinal and Poisonous Plants of Southern and Eastern Africa 2nd Ed Watt JM, Breyer-Brandwijk MG (eds). E & S Livingstone Ltd. Edinburgh. pp 690-695, 1962.
2. Kellerman TS, Coetzer JAW, Naude TW: Plant Poisonings and Mycotoxicoses of Livestock in Southern Africa. Oxford University Press, Cape Town, South Africa. pp 97, 1988.

Brugmansia

Family: Solanaceae

Common Name
Angel's trumpet

Plant Description
Comprising five species and several hybrids, the *Brugmansia* species originate in the Andes Mountains of South America. These evergreen shrubs or small trees have large, soft, elliptic leaves with entire or serrated edges. The characteristic flowers are large, fragrant, pendant, trumpet-shaped, and are white, yellow, or pink in color. The fruits are fleshy, unarmed, globular to long and cylindrical. The leaves and stems when crushed have a strong odor reflective of their alkaloid content (Figures 68-71).

Figure 68 *Brugmansia arborea*

Toxic Principle and Mechanism of Action
All parts of the plant contain significant quantities of tropane alkaloids, including hyoscine (scopolamine), hyoscyamine, and norhyoscine.[1] The flowers may contain as much as 0.83% hyoscine, while the leaves contain 0.4% hyoscine.[2] The flowers of mature, older plants contain as much as

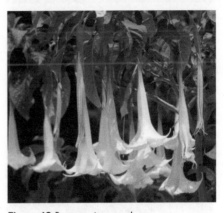

Figure 69 *Brugmansia suaveolens*

3mg of hyoscine.[2] Other plant genera containing similar tropane alkaloids included *Datura* and *Hyoscyamus*. The tropane alkaloids antagonize the actions of acetylcholine and muscarinic, cholinergic receptors, and therefore have a profound, clinically-evident effects upon the autonomic nervous system involving the heart, the digestive system, the eye and the central nervous system. The tropane alkaloids have hallucinogenic properties that have led to human abuse and fatalities.[2,3]

Risk Assessment
Brugmansia species are common garden and patio plants in tropical and subtropical areas and are becoming more common as potted indoor plants in temperate areas. Consequently these plants have good potential for causing poisoning in household pets and children who might eat the leaves and flowers.

Clinical Signs
As with the signs of *Datura* poisoning, the early signs of *Brugmansia* toxicity include depression, unusual behavior, weakness, tachycardia, mydriasis, dry mucous membranes, and constipation. Respiratory failure develops in severe cases.

Figure 70 *Brugmansia sanguinea* hybrid

Figure 71 *Brugmansia* "Double white"

Physostigmine as a cholinergic drug may be used to counter the effects of the tropane alkaloids. However in most cases of human poisoning, conservative supportive treatment without the use of physostigmine generally results in recovery.[4] Seizures can be controlled with the use of diazepam.

References

1. Roses OE, Lopez CM, willGarcia Renadez JC: Isolation and identification of tropane alkaloids in species of the genus *Brugmansia* (Solanaceae). Acta Farm Bonaerense 6: 167-174, 1987.
2. Niess C. Schnabel A. Kauert G. Angel trumpet: a poisonous garden plant as a new addictive drug. Deutsche Medizinische Wochenschrift 124: 1444-1447, 1999.
3. Greene GS, Patterson SG, Warner E: Ingestion of Angel's trumpet: an increasingly common source toxicity. South Med J 89: 365-369, 1996.
4. Rodgers GC, Von Kanel RL: Conservative treatment of jimson weed ingestion. Vet Hum Toxicol 35: 32-33, 1993.

Brunfelsia

Family: Solanaceae

Common Names

Yesterday-today-and tomorrow, lady of the night, Brazil rain tree, noon and night, Paraguayan jasmine

Plant Description

Comprising some 40 tropical American species, these evergreen shrubs or small trees have glossy green, leathery, ovate to elliptic opposite leaves. Flowers are solitary or clustered terminally on the branches and are often fragrant, especially

Figure 72 *Brunfelsia pauciflora*

at night. The 5 sepals are tubular or bell-shaped, 5-toothed, broad lobed; the 5 petals are flat lobed, purple-blue or yellow, fading to white (Figure 72). Fruits are capsules turning brown-black with white juicy flesh, and many small dark seeds.

The most frequently cultivated species of *Brunfelsia* include:

B. americana	Lady of the night
B. australis	Yesterday-today-and-tomorrow
B. pauciflora	Yesterday-today-tomorrow

Toxic principle and Mechanism of Action

All parts of the plant and especially the fruits are toxic, and contain a variety of biologically active compounds. The most toxic of the compounds include brunfelsamidine and hopeanine. Experimentally the compounds both produce convulsions in mice. Brunfelsamidine causes excitement, seizures and death; while hopeanine is more of a depressant and paralytic.[1] Signs of *Brunfelsia* poisoning can mimic those of strychnine poisoning.

Another compound, scopoletin (gelseminic acid), also found in the species *Tetrapleura teraptera*, is a smooth muscle relaxant causing hypotension, through neuromuscular block, and decreased heart rate.[2] This effect appears to be irreversible with atropine.[2] Scopoletin also has strong anti prostaglandin properties.[3]

Risk Assessment

Brunfelsia species are commonly grown in tropical areas as showy ornamental shrubs because of their floriferousness and fragrance. Most reported cases of poisoning have occurred in dogs and cattle after they have eaten the fruit capsules.[4-8]

Clinical Signs

Depending on the amount of the plant or the fruits that have been consumed, dogs show initially excitement, anxiousness, coughing, and vomiting or wretching within in a few minutes to several hours after eating the plant. Muscle tremors, and typical extensor rigidity seizures similar to those of strychnine poisoning then develop.[8] Seizures can be induced by stimuli such as loud noises. Brunfelsia poisoning should be considered in the differential diagnosis of strychnine poisonng. Seizures may persist for several days before the animal recovers. Post mortem findings are minimal and non specific!

Treatment
Treatment of *Brunfelsia* poisoning in dogs is directed at controlling the seizures with diazepam or pentobarbital, along with intravenous supportive therapy during the period of seizuring. Activated charcoal via stomach tube is beneficial, but emetics should be used cautiously as they can instigate seizuring.

References
1. Lloyd HA et al: Brunfelsamidine: a novel convulsant from the medicinal plant *Brunfelsia grandiflora*. Tetrahedon Lifter 26: 2623-2624, 1985.
2. Ojewole JAO, Adesina SK: Cardiovascular and neuromuscular actions of scopoletin from the fruit of *Tetrapleura tetraptera*. Planta Med 49: 99-102, 1983a.
3. Farah MH, Samuelson G: Pharmacologically active phenylpropanoids from *Senra incana*. Planta Medica 58: 14-18, 1992.
4. Tokarnia CH, Dobereiner J, Peixoto PV: Poisonous plants affecting livestock in Brazil.Toxicon 40: 1635-1660, 2002.
5. Spainhour CB, Fiske RA, Flory W, Reagor JC: A toxicological investigation of the garden shrub *Brunfelsia calyina* var floribunda (yesterday-today-and-tomorrow) in three species. J Vet Diag Invest 2: 3-8, 1990.
6. Banton MI, Jowett PL, Renegar KR, Nicholson SS: *Brunfelsia pauciflora* ("yesterday, today, and tomorrow") poisoning in a dog. Vet Hum Toxicol 31: 496-497, 1989.
7. Neilson J, Burren V: Intoxication of two dogs by fruit of *Brunfelsia australis*. Austr Vet J 60: 379-380, 1983.
8. McBarron EJ, de Sarem DW: Letter: poisoning of dogs by the fruits of the garden shrub *Brunfelsia bonodora*. Austr Vet J 51: 280, 1975.

Buxus

Family: Buxaceae

Common Name
Box, boxwood
Commonly cultivated species include
Buxus sempervirens (common or
European box). *B. microphylla* (Japanese
box), and *B. balearica* (Balearic box).

Plant Description
Comprising a genus of approximately 70
species and numerous cultivars,
boxwoods are evergreen branching shrubs
or small trees originating in Europe, Asia,
central America and southern Africa.
Leaves are opposite, short-petiolate,
simple, glossy and leathery (Figure 73).
Numerous small flowers, greenish yellow
in color are produced in clusters in the leaf axils. Fruits are globose to ovoid,
3-horned, leathery capsules containing glossy black seeds.

Figure 73 *Buxus microphylla*

Toxic Principle and Mechanism of Action
Numerous steroidal alkaloids, glycosides, and flavonoids are present in the leaves.[1]
The bitter tasting alkaloids have irritant effects on the gastrointestinal system, causing
excessive salivation colic, profuse diarrhea, and dehydration. Tremors, seizures, and
respiratory difficulty may also occur, suggesting a neurological effect of the toxins.[2]

Risk Assessment
Boxwoods are commonly planted for their attractive evergreen foliage and are
often grown as hedges. Although of relatively low risk to household pets,
boxwoods have caused poisoning in livestock, and have the potential to cause
poisoning in dogs and cats.

Clinical Signs
Excessive salivation, vomiting, abdominal pain, profuse diarrhea, and tenesmus
are typical of the irritating effects of the alkaloids present in boxwoods. Severe
dehydration may result from the diarrhea. Seldom is poisoning fatal, and
treatment should be directed and providing intestinal protections and fluid
therapy as necessary.

References
1. Tomko J, Voticky Z: Steroid alkaloids: the *Veratrum* and the *Buxus* groups. In the
Alkaloids: chemistry and physiology, vol 14. Manske RHF ed. Academic press, New
York pp 1-82, 1973.
2. Kvaltinova Z, Lukovic L, Machova J, Fatranska M: Effect of the steroidal alkaloid
buxaminol-E on blood pressure, acetylcholinesterase activity and (3H) quinuclidinyl
benzilate binding in cerebral cortex. Pharmacology 4: 20-25, 1991.

Family: Fabaceae

Figure 74 *Caesalpinia mexicana*

Figure 75 *Caesalpinia pulcherima*

Common Name

Peacock flower, Barbados pride, bird of paradise flower, dwarf Poinciana Some shrub species of *Caesalpinia* were formally named in the genus *Poinciana* and should not be confused with the "Poinciana" tree of the genus *Delonix*. Three species of *Caesalpinia* are of possible toxicologic significance:[1] *Caesalpinia coriaria, C. gilliesii, C. pulcherima.*

Plant Description

A diverse genus of 70 or more species found in the tropical regions of the world. *Caesalpinia* species are scrambling climbers, shrubs, or trees, often with many thorns. Some species are deciduous in the tropical dry season. Leaves are bipinnate, some very large with numerous leaflets. Flowers are produced as spikes, from the upper leaf axles and are showy, with separate petals, conspicuous stamens in shades of red, yellow, or cream. Fruits are typical leguminous pods (Figures 74 and 75).

Toxic Principle and Mechanism of Action

A variety of toxic compounds are found in the genus *Caesalpinia*, including gallotannins, and diterpenoids. All parts of the plant are toxic, but especially the pods and seeds. The gallotannins our potent irritants, causing severe gastrointestinal disturbances, including vomiting, colic, and severe diarrhea.

Risk Assessment

Caesalpinia species are attractive, shrubs and trees commonly grown as ornamentals in warmer climates. Consequently, domesticated animals can be exposed to the pods and seeds of these plants, at various times of the year. Poisoning has been experimentally produced in animals by feeding dried leaves and seeds.[2,3]

Clinical Signs and Treatment

Vomiting, diarrhea, and in severe cases, dehydration may occur after the ingestion of seeds. Treatment, if necessary, should be directed at relieving the vomiting, diarrhea, and dehydration.

References

1. Burrows GE, Tyrl RJ: Toxic Plants of North America. Iowa State University Press, Ames. pp 537-538, 2001.
2. Connolly JD, Hill RA: Dictionary of Terpenoids vol2. Di- and Higher Terpenoids. Chapman & Hall, London, pp 657-1460, 1991.
3. Watt JM, Breyer-Brandwijk MG: The medicnal and poisonous plants of southern and eastern Africa. E & S Livingston, Edinburgh pp 564-565, 1962.

Caladium

Family: Araceae

Common Name

Caladium, angel's wings, caladio, cananga

Plant Description

Consisting of 7 species of tropical plants from the Americas, the *Caladium* species are frequently grown for their colorful variegated leaves. Growing from underground tubers, the heart-shaped leaves vary considerably in color and markings. The leaves are deciduous in temperate climates. The infloresences are inconspicuous, thin, greenish-white spathes produced under the leaves (Figures 76 and 77).

Figure 76 *Caladium* bicolor

Toxic Principle and Mechanism of Action

Like other members of the arum family, the *Caladiums* contain calcium oxalate crystals in the leaves and stems that can cause inflammation and pain in the mouth of animals or people who chew on the plant parts[1] (see *Dieffenbachia* species).

Risk Assessment

As they are common house and garden plants, caladiums have the potential for causing poisoning in household pets that might chew upon the plant or tubers.

Figure 77 *Caladium* bicolor (inset-flower)

Clinical Signs

Dogs and cats that chew repeatedly on the leaves and stems of caladiums may salivate excessively and vomit as a result of the irritant effects of the calcium oxalate crystals embedded in their oral mucous membranes. The painful swelling in the mouth may prevent the animal from eating for several days. Severe conjunctivitis may result, if plant juices are rubbed in the eye.

Treatment

Unless salvation and vomiting are excessive, treatment is seldom necessary. Anti inflammatory therapy may be necessary in cases where stomatitis is severe. The plant should be removed or made inaccessible to the animals that are eating the plants.

References

1. Genua JM, Hillson CJ: The occurrence, type, and location of calcium oxalate crystals in the leaves of 14 species of Araceae. Ann Bot 56: 351-361, 1985.

Calla pallustris

Family: Araceae

Figure 78 *Calla pallustris*
(With permission from Glen Lee - Saskatchewan Native Plants/Wildflowers website)

Common Names

White arum, water arum, bog arum, wild calla, water dragon

Plant Description

Consisting of a single species, *Calla pallustris* is native to wetland habitats of North America, Europe, and Asia. Growing from a rhizomatous root, it can be a vigorous aquatic plant. The leaves are basal, glossy, green, broadly ovate with acute apices and cordate bases. The inflorescence is a solitary white spathe, with a spadix shorter than the spathe. Fruits are red and enclosed in a gelatinous material (Figure 78).

Calla pallustris should not be confused with calla lilies that are of the genus *Zantedeschia.*

Toxic Principle and Mechanism of Action

Like other members of the Araceae family, *Calla pallustris* contains oxalate crystals in the stems and leaves, and can therefore can cause stomatitis and contact dermatitis.[1] (See *Dieffenbachia* species.)

Risk Assessment

Calla pallustris is an increasingly popular aquatic plant for garden ponds, and consequently household pets have access to the plants.

Clinical Signs

If chewed, excessive salivation and vomit may result from the irritant effects of the calcium oxalate crystals becoming embedded in the oral mucous membranes. The painful swelling in the mouth may prevent the animal from eating for several days. Severe conjunctivitis may result if plant juices are rubbed in the eye.

Treatment

Unless salvation and vomiting are excessive, treatment is seldom necessary. Anti inflammatory therapy may be necessary in cases where stomatitis is severe.

References

1. Franceschi VR, Horner HT: Calcium oxalate crystals in plants. Bot Rev 46: 361-427, 1980.

Calotropis

Family: Asclepiadaceae

Common Name

Crown flower, giant milkweed, king's crown, St. Thomas bush, wild cotton

Plant Description

There are 6 species of *Calotropis* native to northern Africa and south western Asia. Some species are grown as ornamentals in California, Hawaii, Mexico, and Florida. These evergreen shrubs, 6-8ft in height have tough branching stems containing a milky sap. The leaves are opposite, sessile, leathery, oblong, and gray-green in color. Inflorescences are produced in the axils of the upper leaves. Produced in umbels, the flowers have fused sepals and purple to white petals, the corona having 5 fleshy appendages (Figures 79 and 80). The fruits are usually paired pointed follicles with numerous seeds.

Figure 79 *Calotropis gigantea*

Toxic Principle and Mechanism of Action

An array of toxic cardenolides are present in all parts of the plant including the roots.[1-3] These cardiotoxic compounds including calotropin, calactin, uscharidin,

Figure 80 *Calotropis gigantea* flowers

and frugoside are similar to those found in the toxic milk weeds (*Asclepias* species). Preparations from *Calotropis* species have been used for their anti-inflammatory, anthelmintic, larvicidal, irritant and purgative effects in Africa and Asia.[3-6] The latex when fed to goats at a dose of 1.6ml per kg body weight killed the goats within 7 hours.[7] When administered intravenously at a dose of 0.005ml per kg body weight, the latex was fatal to goats within 20 minutes.[8] When dosed orally, the goats developed neurological signs including excessive salivation, frequent urination, nervousness, difficulty in breathing, and diarrhea.[8] The action of the cardenolides is similar to that of ouabain and digitalis and may induce cardiac conduction disturbances, dysrhythmias, and heart block.[9]

Risk Assessment
Since *Calotropis* species are being used as ornamentals in gardens and as house plants especially where drought tolerant plants are desired, young children and household pets can be poisoned if the plants are chewed or eaten.

Clinical Signs
Vomiting, salivation, colic, diarrhea, weakness, heart irregularity, and bradycardia can be anticipated in animals and people as a result of the cardenolides and other irritant compounds present in the plant.

Treatment
Activated charcoal given orally is helpful in preventing absorption of the plant toxins. Symptomatic treatment for colic and diarrhea is indicated in severe cases, as is treatment of the cardiac dysrhythmias (See Aselepias-Milkweeds).

References
1. Hesse G, Heusser LJ, Hutz E, Reicheneder F: African arrow poisons. 5. Relationships between the most important poisons of *Calotropis procera*. Annalen 566: 130-139, 1950.
2. Watt JM, Breyer-Brandwijk MG: The medicinal and poisonous plants of southern and eastern Africa. E & S Livingston, Edinburgh 5119-141, 1962.
3. Radford DJ, Gillies AD, Hinds JA, Duffy P: Naturally occurring cardiac glycosides. Med J Aust. 144: 540-544, 1986.
4. Al-Qarawi AA, Mahmoud OM, Sobaih, Haroun EM, Adam SE: A preliminary study on the anthelmintic activity of *Calotropis procera* latex against Haemonchus contortus infection in Najdi sheep. Vet Res Communications. 25: 61-70, 2001.
5. Kumar VL, Basu N: Anti-inflammatory activity of the latex of *Calotropis procera*. J Ethnopharmacology. 44: 123-125,1994.
6. Mossa JS, Tariq M, Mohsin A, Ageel AM, al-Yahya MA, al-Said MS. Rafatullah S. Pharmacological studies on aerial parts of *Calotropis procera*. American Journal of Chinese Medicine. 19(3-4): 223-231, 1991.
7. el Sheikh HA, et al: The activities of drug-metabolizing enzymes in goats treated orally with the latex of *Calotropis procera* and the influence of dieldrin pretreatment. J Comp Path. 104: 257-268, 1991.
8. el Badwi. Samia MA. Adam SE. Shigidi MT. Hapke HJ: Studies on laticiferous plants: toxic effects in goats of *Calotropis procera* latex given by different routes of administration. DTW - Deutsche Tierarztliche Wochenschrift. 105: 425-427, 1998.
9. Kulkarni SD, Mujumdar SM, Joglekar GA: Effect of *Calotropis gigantea* on dog E.C.G. Indian Heart J. 28: 186-189, 1976.

Caltha
Family: Ranunculaceae

Common Names
Marsh marigold, cowslip, kingcup, bull's eyes

Plant Description
Consisting of 10-20 species native to most temperate regions of the world, the *Caltha* species are colorful wildflowers, some of which have been hybridized into showy ornamentals. Certain cultivars of *Caltha palustris* are prized for their double flowering habit, and are commonly grown around ponds.

Figure 81 *Caltha palustris*

Caltha species are perennials growing from tuberous roots or rhizomes, preferring marshy ground. The leaves are alternate, basal, simple, cordate to reniform, margins entire or toothed, basal leaves having long petioles. Flowers are showy, 5-12 sepals, petals absent, yellow, but can be white to pinkish (Figures 81 and 82).

Toxic Principle and Mechanism of Action
Caltha species contain the irritant glycoside ranunculin that is converted to protoanemonin when the plant tissues are chewed and macerated.[1] Protoanemonin is the vesicant, and it is polymerized to the non-toxic anemonin. The dried plant contains mostly anemonin and is therefore not toxic.

Figure 82 *Caltha palustris*

Risk Assessment
Marsh marigolds are not a significant problem to household pets as the bitter irritant effects of the plants are a deterrent to most dogs and cats.

Clinical Signs
Excessive salivation, vomiting and diarrhea can be anticipated if buttercups are eaten. Treatment if necessary would be symptomatic.

References
1. Hill R, Van Heyningen R: Ranunculin: the precursor of the vesicant substance in Buttercup. Biochem J 49: 332-335, 1951.

Calycanthus
Family: Calycanthaceae

Figure 83 *Calycanthus floridus*

Figure 84 *Calycanthus floridus*

Common Name
Calycanthus floridus - Carolina allspice, sweet shrub
C. *occidentalis* - California allspice, spice bush

Plant Description
Native to North America, the two species of *Calycanthus* are deciduous shrubs grown for their spicy aroma evident when the leaves or stems are cut or bruised. The glossy green leaves have entire margins, and are occasionally serrated. Single aromatic brown to maroon flowers, 4-7cm in diameter are produced amongst the leaves and have a unique strawberry or pineapple scent (Figures 83 and 84). The fruits are cylindrical dark red-brown achenes.

Toxic Principle and Mechanism of Action
Calycanthus species contain a variety of alkaloids, the principle one being calycanthine. It acts like strychnine on the central nervous system depressing acetylcholine depolarization and thereby increasing synaptic transmission with resulting exaggerated neuromuscular reflexes.[1] Calycanthine appears to mediate its convulsant action predominantly by inhibiting the release of the inhibitory neurotransmitter.[2] Muscle rigidity and tetanic seizures occur rapidly after ingestion of the leaves or bark of the plant.

Risk Assessment
Calycanthus species are commonly grown for their desirable scent and unusual attractive flowers. Although the risk of poisoning is low for domestic pets, the plant is sufficiently toxic to be of concern.

Clinical Signs
A sudden onset of hyperesthesia, muscle rigidity, incoordination, tetanic seizures, and protrusion of the third eyelid similar to strychnine poisoning has been reported in cattle.[3] Treatment of affected animals requires the use of drugs such as diazepam or anesthetics to help control seizures. Heavy sedation, a quiet environment, and intravenous fluid therapy should be given as necessary.

References
1. Manske RHF: The alkaloids of the Calycanthaceae. In Alkaloids: Chemistry and Physiology, vol 8. Manske RHF ed. Academic Press, New York, pp 581-589, 1965
2. Chebib M et al: Convulsant actions of calycanthine. Toxicol Appl Pharmacol 190: 58-64, 2003.
3. Bradley RE, Jines TJ: Strychnine-like toxicity of *Calycanthus*. Southwest Vet 14: 71-73, 1963.

Cannabis sativa

Family: Cannabaceae

Common Name
Marijuana, Indian hemp, hashish, ganja, dagga

Plant Description
As a variable single species, *Cannabis* is native to subtropical Asia, and has since been cultivated widely for its value as a fibre producing plant and its hallucinogenic cannabinoid compounds. Erect, branching herbs growing from 1-3m. in height depending on the growing conditions. The plants have a taproot, and produce male and female flowers on different plants. The leaves are palmate, compound, alternate above, opposite below, with 5-11 lanceolate leaflets that are green on top with whitish undersurfaces. Veins on the leaves run to the notches of the serrated edges of the leaves. Inflorescences are produced on short branches at the upper leaf axils. The male and female flowers are different and produced on separate plants. Flowers themselves are small, devoid of petals, with 5 sepals that can be fused or seperate. Fruits are ovoid achenes that are whitish-green with purple mottling (Figures 85 and 86).

Toxic Principle and Mechanism of Action
At least 60 cannabinoid compounds are found in the leaves and flowers of the plant.[1] The most important of the compounds is 9-tetrahydrocannabinol (THC), which is rapidly absorbed from inhaled smoke and less rapidly if ingested. Although unlikely to cause fatalities,

Figure 85 *Cannabis sativa*

Figure 86 *Cannabis sativa*

pets can absorb the THC if they inhale marijuana smoke or if they eat the plant or foods containing the plant.[2-6] Allergic inhalant dermatitis has been reported in a dog that resulted in facial dermatitis, ocular discharge, and pedal pruritis.[7] Other domestic pets including ferrets may be similarly affected.[8] The primary effect of the THC is on the central nervous system where it causes a variety of signs including depression, hallucinations, disorientation, and dizziness. All manner of abnormal behaviors may be seen in intoxicated dogs.

Risk Assessment

In households where *Cannabis* is available and being used by people, there is always the potential for animals to become intoxicated, especially from the deliberate or secondhand exposure to the smoke.

Clinical Signs

Dogs, and other household pets may show a variety of signs including abnormal behaviour, depression, dizziness, hypersensitivity to different stimuli, and muscle tremors. Symptoms may last for up to 24 hours.

Treatment of animals for *Cannabis* intoxication is rarely necessary. In severely affected animals sedation and a quiet enviromnet may be needed. Supportive therapy including activated charcoal orally, emetic drugs, purgatives, and maintaining body warmth are occasionally necessary

References

1. Mason AP, McBay AJ: *Cannabis:* pharmacology and interpretation of effects. J Forensic Sci. 30: 615-63, 1985.
2. Godbold JC Jr. Hawkins BJ. Woodward MG. Acute oral marijuana poisoning in the dog. J Am Vet Med Assoc. 175: 1101-2, 1979.
3. Welshman MD. Doped dobermann. Vet Rec. 119:512, 1986.
4. Godwin RL. Unusual poisoning in a dog Vet Rec. 130: 335-336, 1992.
5. Valentine J. Unusual poisoning in a dog. Vet Rec 130: 307, 1992.
6. Jones DL: A case of canine *Cannabis* ingestion. N Z Vet J. 135-136, 1987.
7. Evans AG. Allergic inhalant dermatitis attributable to marijuana exposure in a dog. J Am Vet Med Assoc 195: 1588-1590, 1989.
8. Smith RA. Coma in a ferret after ingestion of *Cannabis.* Vet Human Toxicol 30: 486, 1988.

Capsicum

Family: Solanaceae

Common Name
Chili pepper, pepper, chili

Plant Description
Consisting of approximately 22 species of tropical annuals or shrubs originating in tropical America, the genus *Capsicum* is universally recognized and grown for its hot, spicy fruits that are central to many of the world's cuisine's.[1] Such spices as cayenne pepper, paprika, chili powder, and Tabasco are derived from the fruits of *Capsicum* species.

Figure 87 *Capsicum annuum*

In general *Capsicum* species are perennial or annual small shrubs with erect, spreading, branched stems and dark glossy green, glabrous leaves with narrow, lanceolate to elliptic blades with entire margins. Inflorescences are either solitary or clusters of star-shaped 5- lobed white, green or purplish flowers. Fruits are berries, glossy, and of various shapes and sizes. Depending on the species and variety of *Capsicum*, the fruits can be extremely irritating to the mucous membranes of the mouth eyes and nose.

Figure 88 *Capsicum annuum*

Most cultivated species and varieties of *Capsicum* originate from 5 species:
   Capsicum annuum - sweet pepper, hot pepper, ornamental pepper, cayenne (Figures 87 and 88)
   C. frutescens - hot pepper, Tabasco pepper, Thai pepper
   C. chinense - habenero pepper (scotch bonnet) (Figures 89 and 90), red savina
   C. baccatum
   C. pubescens

Within *Capsicum annuum* there are 5 main groups:
   Group 1 (Cerasiforme) – ornamental cherry peppers
   Group 2 (Conoides) - ornamental cone-shaped peppers
   Group 3 (Fasiculatum) – ornamental red chili
   Group 4 –(Grossum) – Bell peppers, sweet peppers, green or red wax peppers
   Group 5 (Longum) – spice peppers, cayenne pepper, chili pepper

Figure 89 *Capsicum chinense* (Habenero)

Figure 90 *Capsicum chinense*

Reputedly, the hottest pepper is the tepin (chiltecpin, chiltepin, chile de pajaro) *C.annuum* var. aviculare. The Pequin (chilipiquin, turkey pepper, grove pepper) that grows in Florida, Texas and Mexico is also very "hot". The degree of "hotness"of *Capsicum* species has been quantitated using the Scoville test, and correlates well to the quantity of capsaicinoids in the pepper as determined by high performance liquid chromatography.[2,3]

Toxic Principle and Mechanism of Action

Capsaicin, and to a minor extent 4 other naturally occurring derivatives of capsaicin are the major irritants found in *Capsicum* species. Highest concentrations of the capsaicinoids are found in the fruits and especially in the seeds. Capsaicin appears to block axons by depleting nerve terminals of the neurotransmitter substance P. In so doing, capsaicin causes functional sensory nerve impairment, and has found medicinal use in helping modify chronic pain.[4] Capsaicin also acts on ion channels in cells causing depolarization and release of neurotransmitters which has an overall effect of stabilizing cell membranes.[5,6] Capsaicinoids are being investigated for the treatment of chronic arthritis pain, diabetic neuropathies, cystitis, and human immunodeficiency virus infection.[7,8] High doses of capsaicin can cause intense irritation to the mucous membranes of the mouth, gastrointestinal system, and may cause lethal convulsions.[9]

Risk Assessment

Questions regarding the toxicity of hot peppers are the most frequent calls to poison control centers. This is because of the wide variety of colorful peppers that are sold for cooking and ornamental purposes, especially around holiday occasions, and are therefore readily accessible in many households. Although rarely a problem to household pets the hot peppers are irritants and are capable of affecting dogs and cats that chew and swallow the fruits. Sun dried peppers may be a source of aflatoxins and a variety of fungi.[9]

Clinical Signs

As strong irritants the capsaicinoids cause a burning sensation in the mouth and eyes. Some species of *Capsicum* (Habenero peppers) may cause blistering of the mucous membranes in some individuals. In larger doses vomiting and diarrhea may occur. Experimentally very high doses cause convulsions, and rats fed 10% capsicum developed centrilobular necrosis of the liver.[10] The effects of the capsaicin are usually transitory, but can be relieved by irrigation of the mucous membranes with water

References

1. Bosland PW: Chiles: history, cultivation, and uses. In G. Charalambous (ed.), Spices, herbs, and edible fungi. Elsevier, New York. pp 347-366, 1994.
2. Collins MD, Mayer-Wasmund L, Bosland PW: Improved method for quantifying capsaicinoids in Capsicum using high-performance liquid chromatography. Hort Science 30: 137-13, 1995.
3. Korel F, Bagdatlioglu N, Balaban MO, Hisil Y: Ground red peppers: capsaicinoids content, Scoville scores, and discrimination by an electronic nose. J Agricultural & Food Chemistry 50: 3257-3261, 2000.
4. Cordell GA, Araujo OE: Capsaicin: identification, nomenclature, and pharmacotherapy. Ann Pharmacother 27: 330-336, 1993.
5. Dray A: mechanism of action of capsaicin-like molecules on sensory neurons. Life Sci 51: 1759-1765, 1992.
6. Robbins W: Clinical applications of capsaicinoids. Clinical J Pain 16: 86-89, 2000.
7. Tolan I, Ragoobirsingh D, Morrison EY: The effect of capsaicin on blood glucose, plasma insulin levels and insulin binding in dog models. Phytotherapy Research 15: 391-394, 2000.
8. Glinsukon T et al: Acute toxicity of capsaicin in several animal species. Toxicon 18: 215-220, 1980.
9. Adegoke GO, Allamu AE, Akingbala JO, Akanni AO: Influence of sundrying on the chemical composition, aflatoxin content and fungal counts of two pepper varieties- *Capsicum annum* and *Capsicum frutescens*. Plant Foods for Human Nutrition 49: 113-117, 1996.
10. al-Qarawi AA. Adam SE. Effects of red chilli *(Capsicum frutescens L)* on rats. Vet Human Toxicol 41: 293-295, 1999.

Family: Palmae

Figure 91 *Caryota mitis* leaves

Figure 92 *Caryota mitis* fruits

Common Name

Fishtail palm

Plant Description

A genus of 12 species from tropical Asia and Australasia with unique bipinnate fronds containing the characteristic leaflets similar to a "fishtail". Smaller species have multiple stems that sucker from the base. The common species found in gardens and households, *Caryota mitis*, can attain heights of 9 m (50 ft) under ideal growing conditions. The flowers and fruits are produced from the tops of the stems, the fruits turning dark red, when ripe with three black seeds (Figures 91 and 92).

Toxic Principle and Mechanism of Action

Calcium oxalate crystals (raphides) are contained in specialized cells referred to as idioblasts.[1] Raphides are long needle-like crystals that are bunched together in these specialized cells. When the plant tissue is chewed by an animal, the crystals are extruded into the mouth and mucous membranes of the unfortunate animal. The raphides once embedded in the mucous membranes of the mouth cause an intense irritation and inflammation. The flesh of the fruits contains the greatest amounts of the raphides.

Risk Assessment

Fishtail palms are common household and garden plants, and are therefore readily accessible to household pets. Only pets that chew on plants are likely to be affected by the irritant effects of the fishtail palms. Dermatitis is reported in people who are exposed to the raphides after handling the fruit pulp.[2]

Clinical Signs
Dogs and cats that chew or eat the fruits may develop edema of the oral mucous membranes. Excessive salivation, difficulty in eating, and vomiting may occur. Unless salivation and vomiting are excessive, treatment is seldom necessary. Swelling of the lips and gums may persist for several days. Anti-inflammatory therapy may be necessary in cases where stomatitis is severe. The plant should be removed or made inaccessible to the animals.

References
1. Franceschi VR, Horner HT: Calcium oxalate crystals in plants. Bot Rev 46: 361-427, 1980.
2. Snyder DS. Hatfield GM. Lampe KF. Examination of the itch response from the raphides of the fishtail palm Caryota mitis. Toxicol Appl Pharmac 48: 287-292, 1979.

Cassia
Family: Fabaceae

Common Name
Golden shower tree, Indian laburnum, golden rain tree, monkey-pod tree, purging cassia, Midas tree.

Plant Description
A genus of 100 species, native to tropical southeast Asia, many have been introduced throughout the tropics. Cassia fistula is a tree growing to 10m (30 ft), it is deciduous during the dry or cool season. Leaves are pinnately compound, alternate, with 3-8 pairs of leaflets, blades ovate to elliptic, and waxy white on the lower surface. Inflorescences are showy,

Figure 93 *Cassia fistula*

pendant, axillary racemes to 80 cm in length. Individual flowers are yellow, five petaled, with prominent anthers. The fruits are narrow, hanging cylindrical ponds, up to 2 ft. (60 cm) in length, filled with seeds embedded in a sticky pulp (Figures 93 and 94).

Toxic Principle and Mechanism of Action
The fruits and seeds contain numerous anthraquinones that act as irritants to the digestive tract.[1]

Risk Assessment
As a commonly grown ornamental tree in tropical gardens, dogs and other animals can have access to the seed pods as they fall to the ground. Poisoning however, from the species has not been reported in household pets.

Figure 94 *Cassia fistula* flowers

Clinical Signs

Vomiting, abdominal pain, and diarrhea can be anticipated if large quantities of the ponds and seeds are consumed. Treatment is seldom necessary, and generally consists of symptomatic treatment.

References

1. Burrows GE, Tyrl RJ: Toxic Plants of North America. Iowa State University Press, Ames. 539-541, 2001.

Celastrus

Family: Celastraceae

Figure 95 *Celastrus orbicularis*

Common Name

American bittersweet *Celastrus scandens*
Oriental bittersweet *C. orbiculatus*

Plant Description

Approximately 30 species of *Celastrus* are found in North America, Africa, Australia and Asia, and are closely related to the genus *Euonymus*. The America bittersweet *(Celastrus scandens)* is native to the eastern States and southern Canada. This deciduous, climbing vine that can attain heights of 6-10 m., has alternate, simple, serrated leaves with unisexual flowers. The small greenish-yellow flowers are produced in summer and are followed by bunches of pea-sized fruits that split open to reveal the orange colored inside of the aril and the red seeds (Figures 95 and 96).

Toxic Principle and Mechanism of Action

A variety of sesquiterpene alkaloids including celapanine, celastrine, and paniculatine have been identified in various species of *Celastrus*, but their toxicity is not well established. Cardenolides are also present and may be responsible for cardiac and gastrointestinal disturbances.[1]

Risk Assessment

Poisoning from *Celastrus* species is apparently quite rare despite the appearance of the attractive fruits that often persist on the vine well into the winter after the leaves have fallen. The red fruits are often used for decorating floral arrangements for the winter season and can therefore be potentially hazardous to children and household pets.

Clinical Signs

Vomiting, colic, and diarrhea can be anticipated but are not likely to be severe enough to warrant supportive treatment.

Figure 96 *Celastrus scandens* fruits

References

1. Hart RC: Toxicity of traditional Christmas greens. Indust Med Surg 30: 522-525, 1961.

Cestrum

Family: Solanaceae

Common Name

Day blooming jasmine, wild jasmine, Chinese ink-berry - *Cestrum diurnum*
Night blooming jasmine, poison berry - *C. nocturnum*
Orange cestrum - *C. aurantiacum*
Red cestrum - *C. fasciculatum*
Green cestrum, willow-leaved jasmine - *C. parqui*

Plant Description

The 200 species in the genus *Cestrum* are native to tropical and subtropical America, and the Caribbean islands. Ranging from sprawling shrubs to small trees, *Cestrum* species are mostly evergreen, but can be

Figure 97 *Cestrum aurantiacum*

deciduous in temperate climates. The stems are erect, spreading, and branched. The leaves are lanceolate, elliptic or oval, with entire margins and a strong pungent smell when crushed. Clusters of flowers are produced terminally or in leaf axils. Flowers may be fragrant or not; the calyx having 5 teeth or lobes and the corollas with 5 lobes, are yellow, red, white, or greenish in color depending on the species. The fruits are small berries that are initially green, and ripen to a white, red, black, violet-brown color depending on the species. A variety of hybrids have been developed that are popular in tropical areas (Figures 97-100).

Figure 98 *Cestrum aurantiacum*

Figure 99 *Cestrum aurantiacum* flowers and fruits

Toxic Principle and Mechanism of Action

All plant parts of *Cestrum* species are toxic, especially the berries. Day blooming jasmine *(Cestrum diurnum)* contains a glycoside of 1,25-dihydroxycholecalciferol that is hydrolyzed in the digestive tract to active vitamin D3.[1] Increased levels of vitamin D3 result in the excessive accumulation of calcium in the tissues. Other species of *Cestrum* contain hepatotoxins that cause acute hepatic necrosis and liver failure.[2-4]

Cestrum diurnum when ingested by cattle, sheep, goats, horses, pigs, and poultry causes dystrophic calcification in multiple tissues due to the excessive effects of vitamin D3.[5-7] As a result of the excessive accumulation of calcium, affected animals develop calcification of the tendons, ligaments, and arteries, and as a result, develop a chronic wasting disease characterized by weight loss, stiffness, reluctance to move, and eventually recumbency. Debilitated animals eventually die.

The species of *Cestrum (C. laevigatum, C. parqui)* that are hepatotoxic contain kaurene glycosides such as a carboxyparquin that is a potent hepatotoxin with similarities to carboxy-actractiloside found in cockleburs *(Xanthium species)*.[8] Cattle, sheep, goats and other species eating these plants are susceptible to the hepatotoxins.[2-4] Affected animals develop central lobular necrosis of the liver, and renal tubular necrosis. Other glycoalkaloids and cardenolides present in *Cestrum* species may play a role in the toxicity of the plants.[9,10]

Risk Assessment

In tropical and sub tropical areas, members of the genus *Cestrum* are frequently grown as garden plants, because of their showy flowers, and in the case of *Cestrum nocturnum* its strong, but pleasant nighttime fragrance. Although there are no reports of household pets being poisoned by *Cestrum* species, the fact that they are commonly grown in gardens and occasionally as potted plants, and have attractive white or black berries, makes the plants potentially hazardous to pets and children.

All reported cases of *Cestrum* poisoning have been in cattle, sheep, goats, horses, pigs, and poultry. It is unlikely that other species of domestic animal would eat sufficient quantity of the plants to produce the signs encountered in livestock. Because *Cestrum* species are grown as ornamentals, it is important to avoid poisoning horses and other livestock species through the planting of *Cestrum* in or around livestock enclosures, or by disposing of plant prunings where they are accessible to animals.

Figure 100 *Cestrum diurnum*

Clinical Signs

Animals that have consumed the calcinogenic *Cestrum diurnum* over a period of weeks, develop a syndrome of chronic weight loss, stiffness, reluctance to move, lameness, and eventually become recumbent. Affected animals have elevated blood calcium levels. Death results from progressive calcification of the soft tissues of the body. Horses are often the most noticeably affected.[5] Animals consuming the non-calcinogenic *Cestrum* species develop signs of liver failure, including weight loss, depression, icterus, and hepatic encephalopathy.[2,3] Liver enzymes are typically elevated reflecting the acute hepatic necrosis.

References

1. Hughes MR et al: Presence of 1,25-dihydroxyvitamin D3-glycoside in the calcinogenic plant *Cestrum diurnum*. Nature 268: 347-349, 1977.
2. Van der Lugt JJ, Nel PW, Kitching JP: The pathology of *Cestrum laevigatum* (Schlechtd.) poisoning in cattle. Onderstepoort Vet Res. 58: 211-221, 1991.
3. Van Der Lugt JJ, Nel PW, Kitching JP: Experimentally-induced *Cestrum laevigatum* (Schlechtd.) poisoning in sheep. Onderstepoort J Vet Res 59: 135-144, 1992.
4. McLennan MW, Kelly WR: *Cestrum parqui* (green cestrum) poisoning in cattle. Austr Vet J 1984, 61: 289-291.
5. Krook L et al: Hypercalcemia and calcinosis in Florida horses: implication of the shrub, *Cestrum diurnum*, as the causative agent. Cornell Vet 1975, 65: 26-56.
6. Durand R, Figueredo JM, Mendoza E: Intoxication in cattle from Cestrum diurnum. Vet Human Toxicol 1999, 41: 26-27.
7. Simpson CF, Bruss ML: Ectopic calcification in lambs from feeding the plant Cestrum diurnum. Calcified Tissue International 1979, 29: 245-250.
8. Pearce CM et al: Parquin and carboxyparquin, toxic kaurene glycosides from the shrub Cestrum parqui. J Chem Soc 1992, 1992: 593-600.
9. Mimaki Y. Watanabe K. Sakagami H. Sashida Y. Steroidal glycosides from the leaves of *Cestrum nocturnum*. J Natural Products 2002, 65: 1863-1868.
10. Halim AF. Collins RP. Berigari MS. Alkaloids produced by *Cestrum nocturnum* and *Cestrum diurnum*. Planta Medica 1971, 20: 44-53.

Chelidonium majus
Family: Papaveraceae

Figure 101 *Chelidonium majus*

Figure 102 *Stylophorum diphyllum*

Common Name
Celandine, greater celandine, swallowwort, poppywort, rock poppy

Description
Consisting of a single species originating from Europe and Western Asia, celandine can be found widely cultivated or naturalized in the Eastern regions of North America. It is a short-lived biennial or perennial, growing from rhizomes or taproots to a height of 100 cm, and forming clumps of leafy stems, with alternate, petioled, pinnately divided and toothed leaves. (Figure 101). The plant contains a thick yellow to orange sap that turns red upon exposure to air. The bright yellow 2 sepalled and 4 petalled flowers are produced terminally on the branches or from the leaf axils. Chelidonium majus 'Flore Pleno' has double yellow flowers. The seeds are produced in a narrow oblong capsule and that splits open when ripe to release the seeds.

A similar plant is Stylophorum diphyllum (wood poppy), and is also called celandine poppy by some! (Figure 102).

Toxic Principle and Mechanism of Action
A variety of toxic isoquinoline alkaloids including allocrytapine, berberine, chelidonine, coptisine, protopine, and sanguinarine are found in the sap and other parts of the plant.[1-3] These compounds are bitter tasting, and have spasmolytic effects on the GABA receptors.[2] Primary effects appear to be on the gastrointestinal system with signs of constipation followed by excessive salivation, frequent urination and diarrhea. Affected animals are depressed and drowsy, and may develop seizures in severe cases. The orange latex is irritating to the skin and has been used to treat skin warts. In Europe and China the plant has long been considered an herbal medicine, but greater celandine should be used with caution as it will cause cholestatic hepatitis in people.[4]

The California poppy *(Eschscholzia californica)*, opium poppy *(Papaver somniferum)* and the horned or sea poppy *(Glaucium* species) have similar alkaloids that have the potential to be toxic[5] (Page 205). Blood root *(Sanguinaria Canadensis)* also contains a variety of similar toxic alkaloids including sanguinarine, some of which have antibacterial, anti tumor and cytotoxic activity[6] (Page 243).

Risk Assessment
Animal poisoning is unlikely as celandine has a pungent odor and is likely distasteful. It is however quite accessible to pets as it is grown as a ground cover because it is frost hardy, continuously blooming and self seeding. Similarly blood root *(Sanguinaria Canadensis)* is quite often grown as an ornamental, and the roots with the red sap are potential hazardous to pets that might chew and eat them.

Clinical Signs
Excessive salivation and diarrhea can be anticipated, and may be accompanied by depression and drowsiness. Animals may develop seizures if aroused.[5] Symptoms are generally self limiting, and only rarely is it necessary to treat the animal symptomatically.

References
1. Preininger V: Chemotaxonomy of the Papaveraceae and Fumariaceae. In The Alkaloids, vol 29 Brossi A ed. Academic Press Orlando, Florida, 1-98, 1986.
2. Haberlein H, Tschiersch KP, Boonen G, Hiller KO: *Chelidonium majus* L.: components with in vitro affinity for the GABAa receptor, Positive cooperation of alkaloids. Planta Med 52: 227-231, 1996.
3. Colombo ML, Bosisio E: Pharmacological activities of *Chelidonium majus* L. (Papaveraceae) Pharmacological Research 33: 127-134, 1996.
4. Stickel F, et al: Acute hepatitis induced by Greater Celandine *(Chelidonium majus)*. Scandinavian J Gastroenterology 565-568, 2003.
5. Burrows GE, Tyrl RJ: Toxic Plants of North America. Iowa State University Press, Ames. 846-850, 2001.
6. Kosina P, et al: Sanguinarine and chelerythrine: assessment of safety on pigs in ninety days feeding experiment. Food & Chem Toxicol 42: 85-91, 2004.

Cicuta

Family: Apiaceae

Figure 103 Water hemlock roots

Figure 104 Water hemlock

Common Name
Water hemlock, cowbane, beaver poison, musquash root, poison parsely

Plant Description
Usually occurring as individual or widely dispersed plants along waterways or in marshy ground, the genus *Cicuta*, native to North America, has 4 generally agreed-upon species. These are *Cicuta bulbifera, C. douglasii, C. maculata,* and *C. virosa.*[1,2]

Cicuta species are erect, annual, perennial, or biennial plants, attaining heights of 1-1.5 m (4-5 ft.) The smooth hollow stems arise from thickened tuberous roots, the crowns of which protrude above the soil surface (Figure 103). The hollow stems are partitioned and towards their base the partitions are closer together. The stem base adjacent to the root crown is chambered, the chambers containing a yellowish pungent fluid. The leaves are 1-3 times, pinnately compound; leaflets lanceolate with serrated margins and leaf veins extending to the notches on the serrated margins. The fluorescence is a characteristic open umble, with many white-petaled flowers (Figure 104). The seeds are yellowish brown, oval-shaped with conspicuous ribs.

Toxic Principle and Mechanism of Action
All species of water hemlock and all parts of the plant should be considered poisonous. The fleshy roots contain the highest concentrations of the unsaturated acetylenic alcohols cicutoxin and cicutol. Cicutoxin is highly toxic to all animals, including man, with a lethal toxic dose of the root being 0.5%, body weight or less.[3] Sheep dosed with 2 g per kilogram body weight of the roots died with a 1-2 hours.[3] Cicutoxin acts primarily on the central nervous system, acting in a similar manner to picrotoxin and strychnine by blocking the gamma amino butyric acid (GABA) receptors, and therefore the major inhibitory pathways of the brain.[4] Muscle twitching, followed by seizures, violent chewing movements, opisthotonus, respiratory paralysis, and death occur within 1-2 hours of eating the plant.

Risk Assessment

Water hemlock is rarely intentionally grown as a garden plant. However, the attractive white flowers may encourage enthusiastic gardeners to transplant the water hemlock from the wild into their water garden. Similarly, home owners may have streams, ponds or marshy ground in which the water hemlock may be valued as a natural wildflower on their property. The tuberous roots, green stems and leaves, and the seeds of water hemlock are poisonous. The dried stems appear to be minimally toxic, while the roots remain highly toxic year round. Water hemlock is probably the most poisonous native plant in North America, and therefore warrants extreme caution, where it may be accessible to animals and children. The European *Oenanthe crocata* (water dropwort, dead men's fingers) that closely resembles *Cicuta* species is equally toxic to people and animals.

Clinical Signs

Muscle twitching, excitement, excessive salivation, vigorous chewing movements and teeth grinding, frequent urination precede the onset of seizures. The seizures are frequently violent, causing animals to severely traumatize their tongues. When a lethal dose of water hemlock is consumed, death usually results in 1-8 hours as a result of cardiopulmonary failure. Animals surviving longer than this frequently recover over the next 1-2 days.

There is no specific antidote for treating water hemlock poisoning. If a diagnosis of water hemlock poisoning is recognized early in the course of intoxication, the affected animal should be anesthetized and provided supportive treatment until the seizures diminish or cease. Sheep experimentally poisoned with lethal doses of water hemlock, recovered if they were anesthetized at the onset of clinical signs.[3]

References

1. Mulligan GA: The genus *Cicuta* in North America. Can J Bot 58: 1755-1767, 1979.
2. Burrows GE, Tyrl RJ: Toxic Plants of North America. Iowa State University Press, Ames. pp 49-54, 2001.
3. Panter KE, Baker DC, Kechele PO: Water hemlock *(Cicuta douglasii)* toxicosis in sheep: pathologic description and prevention of lesions and death. J Vet Diagn Invest 8: 474-480, 1996.
4. Wittstock U, Litchnow KH, Teuscher E: Effects of cicutoxin and related polyacetylenes from *Cicuta virosa* on neuronal action potentials: a comparative study on the mechanism of the convulsive action. Planta Med 63: 120-124, 1997.

Clematis

Family: Ranunculaceae

Figure 105 *Clematis jackmanii*

Figure 106 *Clematis* species flower

Common Name
Clematis, virgin's bower, traveller's joy

Plant Description
Consisting of over 200 species, Clematis are common in temperate Asia, Europe, and North America. Clematis species are woody, deciduous or evergreen vines, that climb by means of their twisting leaf-stalk tendrils. The opposite leaves are simple, or compound, with entire or serrated margins. The flowers are produced either singly or in many flowered panicles, the petals are absent, and instead large showy sepals in colors ranging from white, greenish-white, purple, violet, red, or yellow are distinctive (Figures 105 and 106). The many stamens are prominent. The flowers are followed by masses of fluffy, silvery seed heads (Figure 107).

Toxic Principle and Mechanism of Action
Clematis species contain the irritant glycoside ranunculin that is converted to protoanemonin when the plant tissues are chewed and macerated.[1] Protoanemonin levels amongst the species vary, and appear to be low.[2] Protoanemonin is the vesicant, and it is polymerized to the non-toxic anemonin. The dried plant contains mostly anemonin and is therefore not toxic.

Risk Assessment
Clematis are popular garden plants, but are not a significant problem to household pets as the bitter, irritant effects of the plants are a deterrent to most dogs and cats.

Clinical Signs
Excessive salivation, vomiting and diarrhea can be anticipated if clematis are eaten. Treatment in severe cases should include the use of oral activated charcoal (2-8 gm/kg body weight) as an adsorbent, cathartics and intravenous fluid therapy if the animal is severely dehydrated.

References

1. Hill R, Van Heyningen R: Ranunculin: the precursor of the vesicant substance I Buttercup. Biochem J 49: 332-335, 1951.
2. Bonora A, Dall'Olio G, Bruni A: Separation and quantification of protoanemonins in Ranunculaceae by normal and reversed phase HPLC. Planta Med 51: 364-36, 1985.

Figure 107 *Clematis* seed head

Clivia

Family: Liliaceae

Common Name
Kaffir lily

Plant Description
A genus of 4 species of perennial lilies native to southern Africa that are popular garden plants in tropical and sub-tropical areas, and as houseplants in temperate areas. Developing from thick rhizomatous roots, the dark green, long, strap-like leaves are attractive in themselves. The showy flowers are produced terminally on a slightly flattened stems in pale yellow or bright orange-red (Figures 108 and 109). Showy red berry-like fruits are conspicuous.

Figure 108 *Clivia miniata*

Toxic Principle and Mechanism of Action
Several phenanthridine alkaloids including lycorine, clivonine, clivatine, miniatine, and hippeastrine have been identified in the leaves, stems, and bulbs of *Clivia* species.[1,2] The phenanthridine alkaloids are present in many of the Liliaceae, most notably in the *Narcissus* group.

Figure 109 *Clivia miniata alba*

The alkaloids have emetic, hypotensive, and respiratory depressant effects, and cause excessive salivation, abdominal pain, and diarrhea. Calcium oxalate raphides may also contribute to the digestive symptoms.

Risk Assessment
This attractive garden plant, or potted houseplant, has rarely caused poisoning in animals, but it has the potential to do so.

Clinical Signs
Vomiting, excessive salivation, abdominal pain, diarrhea, and difficulty in breathing, are associated with the phenanthridine alkaloids present in the lily family. If large quantities of the leaves and bulb are consumed, depression, ataxia, seizures, and hypotension may develop. Poisoning is rarely fatal, and can generally be treated symptomatically.

References
1. Martin SF: The Amaryllidaceae alkaloids. In The Alkaloids: Chemistry and Physiology. vol 30, Brossi A (ed) Academic Press, San Diego, Calif pp 251-376, 1987.
2. Burrows GE, Tyrl RJ: eds. Liliacea. In Toxic Plants of North America. Iowa State University Press. Ames, Iowa. pp 751-814, 2001

Colchicum
Family: Liliaceae

Figure 110 *Colchicum autumnale*

Common Name
Autumn crocus, meadow saffron, and naked lady. The best known species of the genus is *Colchicum autumnale*, autumn crocus, the stamens of the flowers being the source of saffron, an expensive food coloring and spice.

Plant Description
Comprising a genus of some 60 species, *Colchicum* are native to Europe, North America, and West and Central Asia. Many species originate from the Mediterranean area around Turkey and the Balkans. Arising from corms, almost all species flower before they produce leaves. The leaves are basal, long, linear blades. Flowers are produced singularly or in clusters on top of long tubes, and are generally purple, pink, or white in color. In contrast to *Crocus* species, which have 3 stamens, *Colchicum* flowers each have 6

stamens and three pistils. Many species tend to flower in the fall, but others may flower in the spring (Figures 110 and 111).

Toxic Principle and Mechanism of Action

All parts of the *Colchicum* species are poisonous, the greatest concentrations of toxic alkaloids occurring in the flowers and seeds. The corm is also toxic containing 0.05% alkaloids, in contrast to 0.2% in the seeds. However, the corms are large enough to contain toxic doses of the alkaloids. The primary toxic alkaloid is colchicine, which interferes with cell mitosis. Colchicine and its related alkaloids interfere with microtubular dependent cell function's by binding to tubulin protein, thus blocking mitosis in multiple tissues.[1] All animals are susceptible to the toxic effects of colchicine. Multiple organ failure is characteristic of acute colchicine poisoning. Colchicine also causes marked decreases in prolactin, insulin, glucose tolerance, and catecholamine production. An estimated lethal dose of colchicine is 0.8 mg per kilogram body weight.[2] A lethal dose of the bulbs for calves was 12g/kg body weight.[3]

Figure 111 *Colchicum speciosum*

Another plant that is known to contain colchicine and other similar alkaloids is the Glory Lilly *(Gloriosa superba)* which contains up to 0.36% colchicine in its tubers.[4,5]

Risk Assessment

Colchicum species corms contain sufficient quantities of colchicine to make them one of the more toxic corms or bulbs that can become accessible to household pets. The corms are frequently sold for indoor blooming which entails as little as placing the corm in a saucer with some water on a windowsill. The striking blooms appear in 2-3 weeks. In such circumstances the corms are readily accessible to household pets. Cases of human poisoning occur when the *Colchicum* corm is mistaken for that of wild garlic, or the flowers are consumed.[6,7]

Clinical Signs

The most immediate effect of colchicine poisoning is upon the digestive system, where excessive salivation, vomiting, abdominal pain, and severe hemorrhagic diarrhea are typical.[4] Shortly thereafter, infected animals become weak disoriented, developing seizures, and cardiac irregularities.[2-4] Because of the multisystemic effect of colchicine, death results from severe physiological complications of the cardiovascular, pulmonary, renal, metabolic, and neuromuscular systems. Treatment should be aimed at managing the vomiting, diarrhea, and shock-like signs. Colchicine poisoning in humans has been successfully treated using goat-derived, colchicine-specific Fab fragments.[8]

References

1. Capraro HG, BrossiA,: Tropolonic *Colchicine* alkaloids. In The Alkaloids, Brossi A ed, Academic Press, Orlando, Florida, pp 1-70, 1984.
2. Rochdi M et al: Toxicokinetics of colchicine in humans: analysis of tissue, plasma and urine data in 10 cases. Hum Exp Toxicol 11: 510-516, 1992.
3. Yamada AM et al: Histological study of experimental acute poisoning of cattle by autumn crocus *(Colchicum autumnale)*. Japanese Vet Med Sci 60: 949-952, 1998.
4. Nagaratnam N, DeSilva DPKM, DeSilva N: Colchicine poisoning following ingestion of *Gloriosa superba* tubers. Trop Geogr Med 25: 15-17, 1973.
5. Mendis S: Colchicine cardiotoxicity following ingestion of *Gloriosa superba* tubers. Postgraduate Med J 65: 752-5, 1989.
6. Danel VC, Wiart JF, Hardy GA, Vincent FH, Houdret NM: Self-poisoning with *Colchicum autumnale* L. flowers. J Toxicology - Clinical Toxicol. 39: 409-11, 2001.
7. Gabrscek L. et al: Accidental poisoning with autumn crocus. J Toxicology - Clinical Toxicol. 42: 85-8, 2004.
8. Baud FJ et al: Brief report: treatment of severe colchicine overdose with-colchicines-specific Fab fragments. New Eng J Med 332: 642-645, 1995.

Colocasia

Family: Araceae

Figure 112 *Colocasia* species

Common Name

Elephant's ear, taro, dasheen, kalo, tayo bambou, malanga, cocoyam.

Plant Description

This tropical genus of 6 species is perennial, growing from tuberous roots. The leaves are generally large, heart-shaped, with prominent veins, and tall stems that join the blade a little in from its edge, in contrast to the similar species *Alocasia* and *Xanthosoma* where the petiole joins the leaf at the cleft edge. The flowers are typical of arums, with yellow or cream colored spathes, some species of which are fragrant. The tuberous roots of some species are edible when cooked. *Colocasia* species are often grown as ornamentals for their showy leaves in tropical gardens and as house plants in temperate areas (Figure 112). *Colocasia* species are closely related to *Alocasia* species.

Toxic Principle and Mechanism of Action

Like other members of the Araceae family, *Colocasia* species contain oxalate crystals in the stems and leaves. The calcium oxalate crystals (raphides) are contained in specialized cells referred to as idioblasts.[1-3] Raphides are long needle-like crystals grouped together in these specialized cells, and when the plant tissue is chewed by an animal, the crystals are extruded into the mouth and mucous membranes of the unfortunate animal. The raphides once embedded in the mucous membranes of the mouth cause an intense irritation and inflammation. Evidence exists to suggest that the oxalate crystals act as a means for introducing other toxic compounds from the plant such as prostaglandins, histamine, and proteolytic enzymes that mediate the inflammatory response.[4]

Risk Assessment
As they are common garden plants in tropical areas and as house plants in temperate regions, colocasias have the potential for causing poisoning in household pets that might chew upon the plant.

Clinical Signs
Dogs and cats that chew repeatedly on the leaves and stems of colocasias may salivate excessively and vomit as a result of the irritant effects of the calcium oxalate crystals embedded in their oral mucous membranes. The painful swelling in the mouth may prevent the animal from eating for several days. Severe conjunctivitis may result, if plant juices are rubbed in the eye.

Treatment
Unless salvation and vomiting are excessive, treatment is seldom necessary. Anti inflammatory therapy may be necessary in cases were stomatitis is severe. The plant should be removed or made inaccessible to the animals that are eating the plant.

References
1. Sunell LA, Healy PL: Distribution of calcium oxalate crystals idioblasts in corms of taro (Colocasia esculenta)Am J Bot 66: 1029-1032, 1979.
2. Franceschi VR, Horner HT: Calcium oxalate crystals in plants. Bot Rev 46: 361-427, 1980.
3. Genua JM, Hillson CJ: The occurrence, type, and location of calcium oxalate crystals in the leaves of 14 species of Araceae. Ann Bot 56: 351-361, 1985.
4. Saha BP, Hussain M: A study of the irritating principle of aroids. Indian J Agric Sci 53: 833-836, 1983.

Conium maculatum

Family: Apiaceae

Figure 113 *Conium maculatum*
(Inset-stem showing purple spots)

Common Name
Poison or spotted hemlock, European hemlock, poison parsley, fools parsley,

Plant Description
A biennial introduced to North America from Europe, *Conium maculatum* is an erect highly branched plant with hollow smooth stems covered with purple spots particularly towards the base (Figure 113). The plant has a stout taproot. Leaves are alternate, pinnately compound, fern-like, leaflets oblong to lanceolate that are smooth and hairless. Inflorescences our compound umbels 2-10 cm wide with 8-17 rays. Individual flowers are white with 5 petals and no sepals. Fruits are ovoid shizocarps, flattened and prominently ribbed, turning yellowish-brown when mature. The foliage has a pungent odor similar to parsnip.

Toxic Principle and Mechanism of Action
Several pyridine alkaloids including the highly toxic gamma-coniceine, the precursor of coniine, and N-methylconiine are predominantly responsible for the central nervous system depression and teratogenic effects seen in many species of animal eating the roots, immature vegetation, and especially the seeds.[1,2] Pregnant cattle, sheep, and pigs consuming poison hemlock in the first trimester of pregnancy produce fetuses with cleft palates and variable degrees of limb deformities.[3,4] The alkaloid effects on the central nervous system are poorly understood, but are assumed to be similar to that of nicotine. The alkaloids appear to initially stimulate and then block autonomic ganglia. At high doses, neuromuscular blockade results in death of the animal. A wide variety of animals including cattle, sheep, goats, elk, horses, pigs, poultry, and rabbits have been poisoned by *Conium maculatum*.[5-8]

Risk Assessment
Conium maculatum is unlikely to be a problem to household pets, and no poisoning to date has been reported in the dogs or cats. However, *Conium* is an invasive noxious weed and increasingly finds its way into the environment of property owners with pets. In some instances, it has even been grown as a garden plant.

Clinical Signs
Animals consuming poison hemlock will develop muscle tremors, weakness, in coordination, excessive salivation and lacrimation, increased frequency of urination, and colic depending upon the quantity of plant consumed. In high doses, severe depression, progressive paresis leading to recumbency, bradycardia, and respiratory depression may lead to death of the animal.

In severe cases, treatment consists of activated charcoal orally (2-8gm/kg body weight) to prevent further absorption of the toxic alkaloids, and symptomatic treatment to alleviate clinical signs. Atropine appears to have little effect in reversing the neurotoxicity.

References

1. Lopez TA, Cid MS, Bianchini ML: Biochemistry of hemlock *(Conium maculatum L.)* alkaloids and their acute and chronic toxicity in livestock. A review. Toxicon 37: 841-865, 1999.
2. Panter KE, James LF, Gardner DR: Lupines, poison-hemlock and *Nicotiana* spp: toxicity and teratogenicity in livestock. J Nat Toxins 8: 117-134, 1999.
3. Bunch TD, Panter KE, James LF: Ultrasound studies of the effects of certain poisonous plants on uterine function and fetal development in livestock. J Anim Sci 70: 1639-1643, 1992.
4. Panter KE, Keeler RF, Buck WB: Congenital skeletal malformations induced by maternal ingestion of *Conium maculatum* (poison hemlock) in newborn pigs. Am J Vet Res 46: 2064-2066, 1985.
5. Galey FD, Holstege DM, Fisher EG: Toxicosis in dairy cattle exposed to poison hemlock *(Conium maculatum)* in hay: isolation of Conium alkaloids in plants, hay, and urine. J Vet Diagn Invest 4: 60-64, 1992.
6. Frank AA, Reed WM: Comparative toxicity of coniine, an alkaloid of *Conium maculatum* (poison hemlock), in chickens, quails, and turkeys. Avian Dis 34: 433-437, 1990.
7. Short SB, Edwards WC: Accidental *Conium maculatum* poisoning in the rabbit. Vet Hum Toxicol 31: 54-57, 1989.
8. Panter KE, Keeler RF, Baker DC: Toxicoses in livestock from the hemlocks *(Conium* and *Cicuta* spp.). J Anim Sci 66: 2407-2413, 1998.

Convallaria majalis

Family: Liliaceae

Common Name
Lily of the valley

Plant Description
Convallaria majalis is a popular perennial garden plant originating from Europe and North America. It grows in dense colonies from a slender underground rhizome. Leaves are basal, broadly elliptic to oblong, sheathing, glabrous, and dull green. Inflorescences are terminal one-sided racemes, with 5-18 white flowers, each with 6 fused recurved petals. Some

Figure 114 *Convallaria majalis*

cultivars have pale pink flowers and variegated leaves The flowers are strongly perfumed. Fruits are red berries with numerous seeds (Figures 114 and 115).

Speirantha convallarioides is a very similar perennial plant originating from China, with a similar growth habit, and white, perfumed, star-like flowers produced in a loose cluster.

Figure 115 *Convallaria fruits*

Toxic Principle and Mechanism of Action

Approximately 38 cardenolides have been isolated from *Convallaria*.[1,2] Also present are various saponins. All parts of the plant are poisonous, with the greatest concentration of cardenolides being in the roots. The attractive red berries are the commonest source of poisoning in children. The cardenolides have a digitalis-like activity, causing cardiac conduction disturbances. The plant, *Speirantha convallarioides*, has similar cardenolides to those found in *Convallaria majalis*.[3]

Risk Assessment

Lily of the valley is a common garden plant prized for its perfumed white flowers, and is an attractive groundcover. It can also be grown as a potted plant. The attractive, sweet-tasting, red berries are an attraction to children and not infrequently cause poisoning when ingested. Relatively few cases of animal poisoning from lily of the valley have been reported.[4] The popularity of this plant with its high concentration of cardenolides warrants caution when it is present in gardens frequented by children and household pets.

Clinical Signs

As with digitalis poisoning, clinical signs can vary from vomiting and diarrhea, to cardiac arrhythmias and death.[4] Postmortem findings are usually not specific and a diagnosis is often made by finding the plant parts in the digestive system.

Depending on the time elapsed from when the plant was consumed, induction of vomiting, gastric lavage, or administration of activated charcoal is appropriate. Cathartics may also be used to help eliminate the plant rapidly from the digestive system. Serum potassium levels should be closely monitored and appropriate intravenous fluid therapy initiated as necessary. Phenytoin, as an anti-arrhythmic drug effective against supraventricular and ventricular arrhythmias can be used as necessary. Since the effects of the *Convallaria* cardenolides a very similar to those of digitalis, the use of commercially available digitalis antibody (Digibind – Burroughs Wellcome) may be a beneficial in counteracting the effects on the cardenolides.[5]

References

1. Kopp B, Kubelka W: New cardenolides of *Convallaria majalis*. Planta Med 45: 87-94, 1982a.
2. Kopp B, Kubelka W: New cardenolides of *Convallaria majalis*. Planta Med 45: 195-202, 1982b.
3. Pauli GF: The cardenolides of *Speirantha convallarioides*. Planta Med 61: 162-166, 1995.
4. Moxley RA, Schneider NR, Steinegger DH, Carlson MP: Apparent toxicosis associated with lily-of-the-valley *(Convallaria majalis)* ingestion in a dog. J Am Vet Med Assoc 195: 485-487, 1989.
5. Gfeller RW, Messonier SP: Handbook of small animal toxicology and poisoning 2nd ed. Mosby, St. Louis, Missouri pp 161-162, 2004.

Coriaria

Family: Coriariaceae

Common Name
Coriaria

Plant Description
Low growing shrubs of the Mediterranean area, Europe, Asia, and New Zealand, *Coriaria* species (30) have opposite, entire, lanceolate and prominently-veined leaves. Flowers are small, in erect racemes, 5 sepals and 5 fleshy petals shorter than the sepals. The stigma protrudes from the flower bud and makes the most conspicuous part of the flower. The thickened fleshy petals enclose the true seed. Fruits are pea-sized and berry-like (Figures 116 and 117).

Figure 116 *Coriaria ruscifolia*
Photographer: Art Whistler

Toxic Principle and Mechanism of Action
A toxic glycoside coriamyrtin is present in the fruits and leaves and has been associated with poisoning in people who have eaten the berries.[1,2]

Risk Assessment
Poisoning from *Coriaria* appears to occur mostly in people of the Mediterranean area, although there are reports of some *Coriaria* species being toxic to cattle in New Zealand. Various species are cultivated for their attractive leaves and red, yellow or black fruits and therefore pose a potential risk to pets.

Figure 117 *Coriaria ruscifolia* fruits
Photographer: Art Whistler

Clinical Signs
Signs of poisoning in people include nausea, vomiting, delirium, hallucinations, muscle tremors, and convulsions.[1-3] Treatment is directed toward treating the symptoms and removing the toxic fruits from the digestive system using laxatives and/or emetics.

References
1. Skalli S. David JM. Benkirane R. Zaid A. Soulaymani R. Acute intoxication by redoul *(Coriaria myrtifolia* L.). Three observations. Presse Medicale 31: 1554-1556, 2002.
2. Garcia Martin A. Masvidal Aliberch RM. Bofill Bernaldo AM. Rodriguez Alsina S. Poisoning caused by ingestion of *Coriaria myrtifolia.* Study of 25 cases. Anales Espanoles de Pediatria 19: 366-370, 1983.
3. Lampe KF, McCann MA. AMA Handbook of Poisonous and Injurious Plants. Am Med Assoc, Chicago, Illinois pp 63-64, 1985.

Family: Fumariaceae (Papaveraceae)

Figure 118 *Corydalis ochroleuca*

Figure 119 *Corydalis ochroleuca*

Common Name
Fitweed, corydalis, fumewort, fumeroot, golden smoke

Plant Description
Occurring as annuals or perennials, the genus has approximately 300 species that are native to Asia, Europe, and North America. Plants have erect stems, that may be branched or not, and leaves that are once or twice, pinnately compound, fern-like, hairless or with crystalline vesicles. Inflorescences are racemes or panicles produced terminally or in leaf axils. Flowers are bilaterally symmetrical, tubular with a short backward pointing spur that may be curved. Flower colors range from green, white, yellow, pink to purple and blue depending upon the species (Figures 118-121). Fruits are dehiscent capsules containing from 3 to many kidney-shaped seeds.

Toxic Principle and Mechanism of Action
A variety of toxic isoquinoline alkaloids including apomorphine, cularine, protoberberine alkaloids are present in all parts of the plant.[1] Either individually or in combination the alkaloids have neurologic effects through their antagonistic action on the neurotransmitter gamma amino butyric acid (GABA.)[1,2] The alkaloids cause muscle tremors and seizures depending upon the quantity consumed.

Similar isoquinoline alkaloids are found in the related genera *Dicentra* (bleeding heart), *Papaver* (poppies) and *Fumaria* (fumitory).

Risk Assessment
Poisoning is most likely to occur in livestock grazing the plants. However, *Corydalis* species are becoming more popular as garden ornamentals and therefore they pose a risk to household pets.

Clinical Signs
Muscle tremors, a staggering gait, and an increased respiratory rate may be noticed first. Excessive salivation and regurgitation of ingesta may also occur. Muscle

tremors progress to the point animals are unable to stand, become recumbent, and develop tetanus-like seizures. A marked stiffness to the front legs has been reported. These neurologic episodes may last 30 minutes at which time the animal appears normal. Fatalities may result from the cardiopulmonary depressant effects of the alkaloids.[2,3] Affected animals should be treated with activated charcoal orally to decrease absorption of the alkaloids, and seizuring animals should receive appropriate sedation.

References

1. Preininger V: Chemotaxonomy of Papaveraceae and Fumariaceae. In The Alkaloids, vol 29. Brossi A ed. Academic Press, Orlando, Fla pp 1-98, 1996.
2. Smith RA, Lewis D: Apparent *Corydalis aurea* intoxication in cattle. Vet Hum Toxicol 32: 63-64, 1990.
3. Burrows GE, Tyrl RJ: Toxic Plants of North America. Iowa State University Press, Ames. pp 701-709, 2001.

Figure 120 *Corydalis lutea*

Figure 121 *Corydalis flexuosa*

Corynocarpus

Family: Corynocarpaceae

Figure 122 *Corynocarpus laevigatus*
Photographer: Art Whistler

Common Name
Karaka nut or berry

Plant Description
Native to the tropical areas of New Zealand, Corynocarpus laevigatus is branching tree attaining heights of 15 meters, with large, leathery, glossy, green leaves. Inflorescences are panicles, produced terminally on branches. Flowers are greenish white-yellow. The green fruits ripen to a yellow-orange and are ovoid and 5-7 cm in length (Figure 122).

Toxic Principle and Mechanism of Action
The seeds contain a complex glycoside karakin that is a glucose ester of 3-nitro-proprionic acid (3-NPA).[1,2] An identical toxin is also found in creeping indigo *(Indigofera spicata)*. Karakin exerts its primary toxic effect by interfering with ATP synthesis through inhibition of the TCA cycle enzyme succinate dehydrogenase.[2] The resulting oxidative stress primarily affects the central nervous system. Thorough cooking of the seeds destroys the toxicity and makes them edible. The flesh of the ripe fruits is edible raw. Humans and occasionally dogs have been poisoned by eating the uncooked seeds.

Risk Assessment
Corynocarpus laevigatus is not a common tree in North America and therefore is unlikely to be of concern. However it has been introduced to California, the Gulf coast, and Mexico and can therefore be a local hazard to people and pets.

Clinical Signs
Convulsions and cardiovascular collapse can be expected in animals eating the seeds.[2]

Treatment should be directed at providing symptomatic relief.

References
1. Parton K, Bruere AN, Chambers JR. (Eds): Veterinary Clinical Toxicology. Foundation of Veterinary Continuing Education, Massey University. Palmerston North, New Zealand. 345-347, 2001.
2. Bell ME: Toxicology of karaka kernel, karakin, and beta-nitroproprionic acid. N Z J Sci 17: 327-334, 1974.

Family: Crassulaceae

Common Name
Jade plant, jade tree *(Crassula ovata)*, silver jade plant *(C. arborescens)*.

Similar succulent plants of the Crassulaceae include the *Cotyledon* spp., *Kalanchoe* spp., *Sedum* spp., *Sempervivium* spp. (hen and chicks), and *Echevira* spp. (hen and chicks). The South African plants belonging to the genus *Tylecodon*, are also of the same family.

Figure 123 *Crassula ovata* flowers

Plant Description
Comprising a diverse genus of some 200 species, *Crassula* are native to southern Africa, where they often grow in dry conditions. Annual or perennial, small prostrate plants, shrubs or small, branching trees to 3.5 m (12 ft) in height with succulent, opposite leaves, that may join around the stems. Inflorescences are terminal panicles, cymes of tubular flowers, with 5 petals white, pink, red, or yellow in color (Figures 123-125).

Toxic Principle and Mechanism of Action
Several of the Crassulaceae *(Kalanchoe, Cotyledon, Sempervivium, Tylecodon)* are known to contain toxic bufadienolides with digitalis-like effects on the cardiovascular system.[1-3] Their primary effect is to inhibit Na^+/K^+ adenosine triphosphatase, thereby decreasing the transportation of sodium and potassium across cell membranes which decreases cardiac function. Most cases of poisoning reported in animals have occurred in cattle, sheep, horses, birds, and dogs.[3] Dogs have reportedly been poisoned by eating the meat from goats that have died from eating some of the Crassulaceae.[3] All species of the family Crassulaceae should therefore be considered toxic until proven otherwise.

Figure 124 *Crassula ovata*

Figure 125 *Crassula thyrsoides*

Attempts at experimental intoxication of cats with jade plant was unsuccessful, indicating that this common plant may be of very low toxicity.[4]

Risk Assessment

One of the most frequent calls to poison control centers is regarding the toxicity of jade plant *(C. ovata)* to children and pets. At worst it is listed as potentially toxic, causing vomiting, depression, and ataxia.[5] However, as more of the South African species of the Crassulaceae are imported as house and garden plants, there exists the strong potential that some of the species will be of toxic significance.

Clinical Signs

Vomiting, depression, ataxia, and various digitalis-like cardiac conduction disturbances leading to death may be seen in severely poisoned animals.
The oral administration of activated charcoal shortly after the plant is consumed may help decrease consumption of the toxins. Supportive treatment with fluids and electrolytes to counteract the effects of the persistent diarrhea are indicated. Where cardiac dysrhythmias are severe enough to affect cardiac function, atropine or propranolol may be indicated.

References

1. McKenzie RA, Franke FP, Dunster PJ: The toxicity to cattle and bufadienolide content of six *Bryophyllum* species. Austr Vet J 64: 298-301, 1987.
2. Masvingwe C, Mavenyengwa M: *Kalanchoe lanceolata* poisoning in Brahman cattle in Zimbabwe: the first field outbreak. J South African Vet Assoc 68: 18-20, 1997.
3. Kellerman TS, Coetzer JAW, Naude TW. Plant Poisonings and Mycotoxicoses of Livestock in Southern Africa. Oxford University Press, Capetown pp 99-108 , 1990.
4. Burrows GE, Tyrl RJ. Toxic Plants of North America. Iowa State University Press, Ames 2001, pp 389-390.
5. Household Plant Reference. ASPCA National Animal Poison Control Center. 424 E. 92nd St., New York, NY 10128.

Crinum

Family: Liliaceae

Figure 126 *Crinum asiaticum*

Common Name

Crinum lily, swamp lily, Asiatic poison lily, Cape lily, Murray lily

Plant Description

Consisting of about 130 species, these large lilies are closely related to the *Amaryllis* species, and are native to tropical areas around the world. The long, dark green, strap-like leaves arise from a large bulb that has a neck formed by the remnants of the dried leaves and stems. The flowers open progressively as umbels at the ends of long flower stalks. Colors vary from white to pink (Figures 126 and 127). Petals are 6, broad and upward curving. Stamens are held at the ends of long filaments. Fruits are globular and contain large fleshy seeds that require no dormant period for growth.

Toxic Principle and Mechanism of Action

Several phenanthridine alkaloids including lycorine, crinamine, and haemanthamine have been identified in the leaves, stems, and bulbs of *Crinum* species.[1,2] The phenanthridine alkaloids are present in many of the Liliaceae, most notably in the *Narcissus* group. The alkaloids have emetic, hypotensive, and respiratory depressant effects, and cause excessive salivation, abdominal pain, and diarrhea. Calcium oxalate raphides may also contribute to the digestive symptoms.

Figure 127 *Crinum asiaticum* (pink)

Risk Assessment

This attractive garden plant, and occasionally potted houseplant, has rarely caused poisoning in animals, but it has the potential to do so.

Clinical Signs

Vomiting, excessive salivation, abdominal pain, diarrhea, and difficulty in breathing, are associated with the phenanthridine alkaloids present in the lily family. If large quantities of the leaves and bulb are consumed, depression, ataxia, seizures, and hypotension may develop. Poisoning is rarely fatal, and can generally be treated symptomatically.

References

1. Martin SF: The Amaryllidaceae alkaloids. In The Alkaloids: Chemistry and Physiology. vol 30, Brossi A (ed) Academic Press, San Diego, Calif 251-376, 1987.
2. Burrows GE, Tyrl RJ: eds. Liliacea. In Toxic Plants of North America. Iowa State University Press. Ames, Iowa. pp 751-814, 2001

Cryptostegia
Family: Asclepiadaceae

Figure 128 *Cryptostegia grandiflora*
Photographer: Brimiley Burbidge

Common Name
Rubber vine, pink alamanda, palay rubber vine

Plant Description
Consisting of 2 species, *Crypotostegia* originate in tropical Africa, and Madagascar; *Cryptostegia grandiflora* and *C. madagascariensis.*
Mostly small shrubs or woody climbers, with long climbing stems. The sap is milky. Leaves are opposite, petiolate, elliptic, thick and glossy dark green. The flowers are large and showy, bell or funnel-shaped, pink-lavender in color, the underside of the petals darker lavender. Fruits are paired ovoid follicles that contain many tufted, wind born seeds (Figure 128).

A plant called mandevilla (*Mandevilla* species) with similar flowers, and also known commonly as pink alamanda, is not related to the genera *Cryptostegia* or *Allamanda,* and is not known to be toxic.[1]

Toxic Principle and Mechanism of Action
Cryptostegia species contain significant quantities of the cardenolides digitoxigenin, oleandrigenin, and cryptograndosides A & B.[2-4] Saponins have also been isolated from the plant.[3] The high concentration of cardenolides have cardiotoxic and digestive effects much like the milkweeds and other members of the Asclepiadaceae. Similar immunoreactive cardiac glycosides as detectable by radioimmumoassay using antibodies to cardiac glycosides are detectable in other plants including *Asclepias, Nerium, Thevetia, Ackocanthera,* and *Calotropis,* species.[4]

Risk Assessment
Both species of *Cryptostegia* are commonly grown in tropical areas *as* attractive garden plants, and as house plants in temperate areas. Although there are no reports of poisoning in household pets associated with the plant, its significant cardenolide content warrants it be considered a potentially toxic plant. The only reported cases of poisoning have been in horses and cattle.[3,5] The lethal dose in the horse has been estimated to be 0.03-0.06% body weight of the green leaves.[6]

Clinical Signs
Cardenolide poisoning typically is associated with abdominal pain, diarrhea, anorexia, depression, bradycardia, dyspnea, weakness, collapse and death in severe cases. Depending upon the severity of clinical signs, activated charcoal orally, and treatment for bradycardia and dehydration from diarrhea may be necessary.

References
1. Whistler WA: Tropical Ornamentals: a guide. Timber Press, Portland, Oregon. pp 164-165, 2000.
2. Sanduja R, Lo WYR, Euler KL, Alam M: Cardenolides of *Cryptostegia madagascariensis*. J Nat Prod 47: 260-265, 1984.
3. Morton JF: Plants Poisonous to People in Florida. 2nd ed. JF Morton ed. Southeastern Printing, Stuart, Florida pp 9-10, 1982.
4. Radford DJ, Cheung K, Urech R, Gollogly JR, Duffy P: Immunologic detection of cardiac glycosides in plants. Austr Vet J 71: 236-238, 1994.
5. Cook Dr, CampbellGW, Meldrum AR: Suspected *Cryptostegia grandiflora* (rubber vine) poisoning in horses. Austr Vet J 1990, 67: 344, 1990.
6. Everist SL: Poisonous Plants of Australia. 2nd ed. Angus & Roberston, Sydney, Australia pp 94-109, 1981.

Cycas

Family: Cycadaceae

Common Name
Cycad, sago palm, fern palm

Plant Description
Considered to be some of the most primitive of plants, members of the genus *Cycas* are native to the tropical areas of Eastern Africa, Madagascar, northern Australia and many Pacific islands. There are approximately 45 species of *Cycas*, 2 of which are commonly grown as ornamentals: *Cycas revoluta* (sago palm), and *C. circinalis* (fern palm, queen sago). *Cycas* are closely related to the *Macrozamia* species from Australia, and the *Zamia* species from Florida and the Caribbean area.

Figure 129 *Cycas revoluta*

Plants are tree-like, perennial, evergreen, and fern-like, but are not ferns from which they differ substantially. The large (2-3 meters long), pinately compound leaves are spirally clustered at the tops of the thick stems (Figures 129 and 130). Leaflets are coiled when in bud form, and linear to lanceolate or curved, leathery and with the sheathing petioles. Male and female reproductive organs are produced on different plants, the male erect, cylindrical, pollen cones are formed terminally and have motile pollen; the female seed cones ring the stem apex, and may droop like a skirt once the fruits mature. The fruits consist of numerous flattened, fleshy seeds that vary from deep red to orange in color.

Figure 130 *Cycas* species (male)

Toxic Principle and Mechanism of Action

Cycas species contain glycosides of methylazoxymethanol (MAM) and an amino acid beta-N-methylamino-L-alanine (BMAA).[1] The glycosides are found in highest concentration in the leaves, and at lower concentrations in the seeds and pith of the trunks. The primary MAM glycosides cycasin and macrozamin are readily hydrolysed by the plant glycosidases or digestive bacterial enzymesto the toxic aglycones which alkylate DNA and RNA causing cell necrosis. The primary effect is one of hepatic necrosis. However, the alkylating effect on DNA and RNA can also result in the aglycones being carcinogenic, mutagenic, teratogenic, and neurotoxic.[2] The toxic amino acid BMAA is found predominantly in the seeds and causes neurologic lesions similar to other neuro-lathyrogens. Prolonged consumption of the *Cycas* seeds or flour made from the seeds is necessary to induce neurologic signs and lesions.[3]

All animals including people, cattle, goats, sheep, and dogs that eat the fruits, starchy stems, or the leaves are susceptible to poisoning.[4-8]

Risk Assessment

Cycas species are popular garden plants in tropical areas and are often grown as specimen plants for their attractive foliage. In temperate climates the plants are often sold as potted plants for in-door use. The ripe fruits are particularly enticing to dogs and people.[5,10] The fruits of the *Zamia* and *Macrozamia* species are also toxic.[10] Most poisoning in humans occurs when the fruits are misused as a food source, or as alternative medicines for various ailments.[4] The fruits and seeds of some cycads remain toxic even after they are cooked.

Clinical Signs

The primary effect of the *Cycas* toxins is on the liver and digestive tract, with about half the cases of dogs with cycas poisoning developing neurologic signs.[5] Vomiting starts within a few minutes of ingesting the seeds and may persist for hours. Excessive salivation and increased thirst is often evident. During the next few days anorexia, depression, diarrhea or constipation, and icterus develop. Serum bilirubin, liver enzymes, blood urea nitrogen, and creatinine levels increase. The prognosis is guarded to poor once evidence of hepatic necrosis develops. Neurological signs are more common in people and livestock who have consumed *Cycas* plants over a prolonged period.[8,11]

References

1. Yagi F, Tadera K: Azoxyglycoside contents in seeds of several cycad species and various parts of the Japanese cycad. Agric Biol Chem 51: 1719-1721, 1987.
2. Newberne PM. Biologic effects of plant toxins and aflatoxins in rats. J National Cancer Institute 56: 551-555, 1976.
3. Duncan MW, Kopin IJ, Crowley JS, Jones SM, Markey SP: Quantification of the putative neurotoxin 2-amino-3(methylamino)-propanoic acid (BMMA) in Cycadales: analysis of the seeds of some of the members of the family Cycadaceae. J Anal Toxicol 13(4): suppl A-G, 1989.

4. Chang SS, Chan YL, Wu ML, Deng JF, Chiu T, Chen JC, Wang FL: Tseng CP. Acute Cycas seed poisoning in Taiwan. J Toxicology - Clinical Toxicol. 42: 49-54, 2004.
5. Albertson JC, Khan SA, Richardson JA: Cycad palm toxicosis in dogs: 60 cases (1987-1997). J Am Vet Med Assoc 213: 99-101, 1998.
6. Senior DF et al: Cycad poisoning in the dog. J Anim Hosp Assoc 21: 103-109, 1985.
7. Gabbedy BJ, Meyer EP, Dickson J: Zamia palm *(Macrozamia reidlei)* poisoning of sheep. Aust Vet J 51:303-305, 1975.
8. Hall WTK, McGavin MD: Clinical and neurological changes in cattle eating the leaves of *Macrozamia lucida* and *Bowenia serrulata* (family Zamiaceae). Pathol Vet 5: 26-34, 1968.
9. Botha CJ, Naude TW, Swan GE, Ashton MM, van der Wateren JF: Suspected cycad *(Cycas revoluta)* intoxication in dogs. J South African Vet Assoc 62: 189-190, 1991.
10. Mills JN, Lawley MJ, Thomas J: Macrozamia toxicosis in a dog. Austr Vet J 73: 69-72, 1976.
11. Hooper PT, Best SM, Campbell A: Axonal dystrophy in the spinal cords of cattle consuming the cycad palm, *Cycas media.* Austr Vet J 50: 146-148, 1974.

Cyclamen

Family: Primulaceae

Common Name
Cyclamen, Persian violet, sow bread

Plant Description
A genus of about 20 species native to the Mediterranean area, *Cyclamen* species have become popular house plants (*Cyclamen persicum* and its hybrids). Developing from round tubers that sit close to the soil surface, the leaves are basal on variable length petioles, heart or kidney shaped, variegated in shades of

Figure 131 *Cyclamen persicum* cultivar

green and silver-grey. Flowers are solitary, nodding, although the 5 petals are sharply reflexed and erect. Flower color varies from white, pink, to red (Figures 131-133). Fruits are dehiscent capsules with many seeds.

Toxic Principle and Mechanism of Action
All parts of the plant, but especially the tubers contain irritating terpenoid saponins, mainly glycosides of the cyclamiritens and cyclamigenins.[1,2] The saponins have cardiotoxic potential, and because of their irritating properties cause gastrointestinal problems including salivation, vomiting, colic, and diarrhea.

Figure 132 *Cyclamen persicum* flowers

Figure 133 *Cyclamen persicum* hybrid

Risk Assessment
Cyclamens are very common potted house plants, and in milder climates are often successfully grown in rock gardens. The tubers have the potential to be a problem to household pets, but the risk is minimal.

Clinical Signs
Vomiting, abdominal pain, and diarrhea are the most common manifestations of poisoning. Cardiac dysrhythmias and seizures may be seen where higher doses of the plant are consumed. Supportive treatment for the vomiting and diarrhea, when necessary, is all that is generally required.

References
1. Spoerke DG, Spoerke SE, Hall A. Rumack BH: Toxicity of *Cyclamen* persium (Mill). Vet Human Toxicol 29: 250-251, 1985.
2. Spoerke DG, Smolinske SC: Toxicity of House Plants. CRC Press. Boca Raton, Florida. pp 113-114, 1990.

Daphne
Family: Thymelaeaceae

Figure 134 *Daphne burkwoodii*

Common Name
Daphne, spurge laurel, mezereon *Daphne* of toxicologic significance include *D. cneorum, C. genkwa, D. laureola,* and *D. odora.*[1]

Plant Description
Comprising a genus of some 50 or more species, *Daphne* are native to Europe, North Africa, and subtropical Asia. Deciduous or evergreen, erect, woody, branching, shrubs with alternate or opposite, glossy green, ovate to lanceolate leaves. Inflorescences are terminal or axillary, single or clusters of 4 lobed, fragrant flowers that are white, greenish-

white, pink, purple or yellow-orange in color depending on the species. Many cultivars have been developed that have enhanced flower color, and some have variegated leaves. Fruits are globular to ovoid, leathery or fleshy, yellow, red, orange or black drupes (Figures 134-136).

Toxic Principle and Mechanism of Action

The leaves and fruits of *Daphne* contain a variety of bitter tasting tricyclic daphnane and tigliane diterpenes.[2] Poisoning has occurred in children eating the attractive berries, and in livestock browsing on the plants.[1] In addition to the irritant effects on the digestive tract, seizures, tremors and deaths have been reported in children who ate the fruits.[1,3]

Figure 135 *Daphne burkwoodii* 'Carol Mackii'

Risk Assessment

Daphne are commonly grown as garden shrubs for their attractive foliage, flowers, and fruits. The colorful fruits pose the greatest risk to children or animals that eat them. The bitter taste of the fruits generally limits intake and therefore the severity of poisoning.

Figure 136 *Daphne mezereum* leaves / fruits

Clinical Signs

Intense reddening and swelling of the oral mucous membranes, excessive salivation, blistering of the tongue and lips, and vomiting are common effects of *Daphne* poisoning. Diarrhea with blood may occur if sufficient plant material was swallowed. Treatment usually requires activated charcoal orally, along with fluid therapy where diarrhea leads to dehydration.

References

1. Kingsbury JM. Poisonous Plants of the United States and Canada. Prentice-Hall Inc. Englewood Cliffs New Jersey 386-388, 1964.
2. Connolly JD, Hill RA: Dictionary of Terpenoids vol 2. Di- and Higher terpenoids. Chapman & Hall, London 655-1460, 1991.
3. Habermehl GG: Poisonous plants of Europe. In Poisonous Plants: Proceedings of the Third International Symposium. James LF, Keller RF, Bailey EM, Cheeke PR, Hegarty MP eds. Iowa State Univ Press, Ames. 74-83, 1992.

Family: Solanaceae

Figure 137 *Datura metel florepleno*

Figure 138 *Datura wrightii*

Common Name
Moon flower, jimson weed, sacred datura, angel's trumpet, thorn apple, Indian apple, tolguacha.

The 7 North American species include: *Datura discolor, D. ferox, D,inoxia (D. meteloides), D. metel (D. fastuosa), D. quercifolia* (oak-leaf thorn apple), *D. stramonium (Jimson weed), D. wrightii.*(sacred datura).[1]

The genus *Brugmansia* (Angel's trumpet) is very similar to *Datura*, and equally as toxic. (see *Brugmansia*).

Plant Description
Consisting of some 25 species from the tropical and subtropical regions of the world, *Datura* species are shrubby herbs, annuals, or perennials with glabrous or pubescent stems and leaves. The stems are erect, branching, and up to 4 ft. (1.5m) in height. The leaves have short petioles, ovate, elliptic or triangular blades and entire or coarsely toothed or lobed margins and an unpleasant pungent odor when crushed. Large showy fragrant short-lived flowers are produced at the leaf axils. The calyces are tubular and 5-toothed. The corollas are a radially symmetrical, 5-10 lobed, and generally white, yellow, or violet-purple. The fruits are ovoid spiny capsules that split open when ripe to release numerous brown to black seeds with a pitted surface (Figures 137 and 138).

Toxic Principle and Mechanism of Action
All parts of the plants and especially the seeds contain the tropane alkaloids L-hyoscyamine and scopolamine (L-hyoscine). Racemization of hyoscyamine into its D and L forms produces atropine. Hyoscyamine tends to be concentrated in the seeds, while scopolamine is prevalent in the leaves. The concentration of hyoscyamine in the seeds of *Datura stramonium* can reach concentrations of 0.2-0.6%.[2] The tropane alkaloids are acetylcholine antagonists acting at the muscarinic cholinergic receptors in the autonomic nervous system, central nervous system, heart, and digestive systems.

Most cases of poisoning are reported in humans who deliberately consume the seeds or make tea from the leaves to experience the hallucinogenic properties of the tropane alkaloids.[3,4] Although the alkaloids are hallucinogenic, the profound effects of the alkaloids on the nervous system can and do cause fatalities. Honey made from bees that feed predominantly on *Datura* is also toxic.[5] Poisoning in ruminants, horses, pigs and birds generally occurs when *Datura* seeds contaminate the grain they are fed.[6-8] Accidental poisoning occasionally occurs in dogs.[9]

Risk Assessment

Various species of *Datura* are commonly grown as garden plants for their showy display of fragrant white trumpet-shaped flowers. The leaves have a strong odor which generally deters consumption of the plant. However, when the seed capsules ripen and release their seeds, animals can have access to them, and if they are chewed and swallowed can cause severe poisoning. Using the seed pods for dry flower arrangements can increase the potential for poisoning of household pets, including pet birds, unless care is taken to first remove the seeds from the pods.

Clinical Signs

Dilated pupils, decreased salivation, anorexia, intestinal stasis and bloating, constipation, and an increase in respirations and heart rate are typical of the effects of the tropane alkaloids. The neurologic effects, especially the euphoria and seizures, are a feature of human poisoning.

Treatment is usually symptomatic and conservative. Activated charcoal orally as an adsorbent may reduce further absorption of the tropane alkaloids. Fluids and electrolytes intravenously may be necessary in severely intoxicated animals. Seizures may be controlled by diazepam, and in severe cases, physostigmine may be used as a short acting cholinergic to reverse the atropine effects on the autonomic nervous system.

References

1. Burrows GE, Tyrl RJ: Toxic Plants of North America. Iowa State University Press, Ames. 1113-1118, 2001.
2. Friedman M, Levin CE: Composition of jimson weed *(Datura stramonium)* seeds. J Agric Food Chem 37: 998-1005, 1989.
3. Boumba VA, Mitselou A, Vougiouklakis T: Fatal poisoning from ingestion of *Datura stramonium* seeds.Vet Human Toxicol 46: 81-82, 2004.
4. Salen P, Shih R, Sierzenski P, Reed J: Effect of physostigmine and gastric lavage in a *Datura stramonium*-induced anticholinergic poisoning epidemic. Am J Emerg Med 21: 316-317, 2003.
5. Ramirez M, Rivera E, Ereu C: Fifteen cases of atropine poisoning after honey ingestion. Vet Human Toxicol 41: 19-20, 1999.
6. El Dirdiri NI, Wasfi IA, Adam SEI, Edds GT: Toxicity of *Datura stramonium* to sheep and goats. Vet Hum Toxicol 23: 241-246, 1981.
7. Nelson PD, Mercer HD, Essig HW, Minyard JP: Jimson weed toxicity in cattle. Vet Hum Toxicol 24: 321-325, 1982.
8. Schulman ML, Bolton LA: Datura seed intoxication in two horses. J South African Vet Assoc 69: 27-29, 1998.
9. Tostes RA: Accidental *Datura stramonium* poisoning in a dog. Vet Human Toxicol 44: 33-34, 2002.

Family: Ranunculaceae

Figure 139 *Delphinium elatum* hybrid

Figure 140 *Delphinium elatum* hybrid flowers

Common Name

Larkspur, delphinium

Plant Description

The genus *Delphinium* consists of some 250 species native to the Northern Hemisphere, with only a few species found in Africa. Numerous cultivars and hybrids have been developed as popular ornamentals. There are 61 species of Delphinium native to North America.[1] *Delphiniums* are annual or perennial herbs developing from a woody root or rhizome. Stems are erect and hollow. Leaves are simple, alternate, basal, the basal leaves generally being larger than the leaves higher on stems. The blades are deeply palmately divided into 3-7 major lobes, each lobe being further divided. Inflorescences of terminal racemes or occasionally panicles. Flowers are perfect, bilaterally symmetrical, with 5 sepals, 4 petals with the two upper forming a char-acteristics spur (Figures 139-141). Nectaries are present in the spur. Fruits of follicles with curved beaks and numerous dark brown seeds.

The annual larkspurs are now placed in the genus *Consolida*. The showy garden *Delphiniums* have been mainly derived from *Delphinium elatum*, and not the more toxic native species such as *Delphinium barbeyi, D. californicum, D. glaucescens, D. geyeri,* and *D. nuttallianum.* It is of interest to note that larkspur poisoning causes more cattle deaths in the Rocky Mountain region than any other genus of plants[2,3].

Toxic Principle and Mechanism of Action

Delphinium species contain a large number of diterpenoid alkaloids, with some of the more toxic species such as *Delphinium nuttallianum* containing some 27 alkaloids.[4-8] The 19-C diterpene alkaloids including methyllycaconitine, nudicauline, and 14-deactylnudicauline have been shown to have the greatest toxicity in animals. These alkaloids have a curare-like, competitive, non depolarizing, neuromuscular blocking effect by inhibiting acetylcholine at the nicotinic postsynaptic receptor sites.[5] The levels of alkaloid vary considerably with the species and the stage maturity of the plant. The preflowering plant has the highest concentration of alkaloids.[10,11] Cattle are the most susceptible to the toxic effects of the alkaloids, while sheep are quite resistant. Horses are susceptible to poisoning, but rarely eat the plants. Mice are quite susceptible to poisoning and are a good experimental model species.[12]

Risk Assessment
Larkspurs are commonly grown as colorful garden perennials, but are unlikely to be poisonous to household pets. However, because of their potential for poisoning livestock species, care should be taken in disposing of the plants from the garden.

Figure 141 *Delphinium virescens* flower

Clinical Signs
The neuromuscular blocking effect of the toxic *Delphinium* species develops three to four hours after the ingestion of the plant. Initially, muscle tremors, weakness, and an inability to stand are common. Lateral recumbency, bloating, and death occur if a lethal dose of the alkaloids is consumed.

Physostigmine at a dose of 0.04-0.08mg per kilogram body weight given inter peritoneally is the most effective means of reversing the neuromuscular blockade (intravenous administration should be done very cautiously). In cattle it is important to keep the animals in sternal recumbency to avoid bloat.

References
1. Warnock MJ: *Delphinium.* In flora of North America, vol 3 Magnoliophyta: *Magnoliidae* and *Hamamelidae*, Flora of North America Editorial Committee eds, Oxford University Press, New York pp 196-240, 1997.
2. Pfister JA, Gardner DR, Stegelmeier BL, Hackett K, Secrist G: Catastrophic cattle loss to low larkspur *(Delphinium nuttallianum)* in Idaho. Vet Hum Toxicol 45: 137-139, 2003.
3. Pfister JA, Gardner DR, Panter KE, Manners GD, Ralphs MH: Stegelmeier BL. Schoch TK. Larkspur (*Delphinium* spp.) poisoning in livestock. J Nat Toxins 8: 81-94, 199.
4. Bai Y, BennMH, Majak W: The alkaloids of *Delphinium nuttallianum:* the cattle poisoning low larkspur of interior British Columbia. In Poisonous Plants: proceedings of the third international symposium, James LF, Keeler RF, Bailey EM, Cheeke PR, Hegarty MP eds. Iowa State University press, Ames, pp 304-308, 1992.
5. Dobelis P, Madl JE, Pfister JA, Manners GD, Walrond JP: Effects of Delphinium alkaloids on neuromuscular transmission. J Pharmacol Exp Therapeutics 291: 538-46, 199.

6. Stegelmeier BL, Hall JO, Gardner DR, Panter KE: The toxicity and kinetics of larkspur alkaloid, methyllycaconitine, in mice. J Anim Sci 81: 1237-1241, 2003.
7. Batbayar N, et al: Norditerpenoid alkaloids from *Delphinium* species. Phytochemistry. 62(4): 543-50, 2003.
8. Gardner DR, Manners GD, Panter KE, Lee ST, Pfister JA: Three new toxic norditerpenoid alkaloids from the low larkspur *Delphinium nuttallianum*. [erratum appears in J Nat Prod 2000, 63:1598]. J Nat Products 2000, 63: 1127-1130.
9. Nation PN, Ben MH, Roth SH, Wilkens JL: Clinical signs and studies of the site of action of purified larkspur alkaloid, methyllycaconitine, administered parenterally to calves. Can Vet J 1982, 23: 264-266.
10. Ralphs MH, Gardner DR: Distribution of norditerpene alkaloids in tall larkspur plant parts through the growing season. J Chem Ecol 2003, 29: 2013-2021.
11. Ralphs MH, Gardner DR, Turner DL, Pfister JA, Thacker E: Predicting toxicity of tall larkspur *(Delphinium barbeyi)*: measurement of the variation in alkaloid concentration among plants and among years. J Chem Ecol 2002, 28: 2327-2341.
12. Olsen JD, Sisson DV: Toxicity of extracts of tall larkspur *(Delphinium barbeyi)* in mice, hamsters, rats and sheep.Toxicology Letters 1991, 56: 33-41.

Dicentra

Family: Fumariaceae

Figure 142 *Dicentra spectabilis*

Common Name
Bleeding heart, squirrel corn, Dutchman's breeches

Plant Description
A genus of approximately 20 species of annuals and perennials, native to North America and Asia, *Dicentra* species are popular ornamentals. Plants are either stemless or have stems with distinct nodes, growing to 2 ft. (50cm) in height, with once or twice pinnately compound, glabrous or glaucous, fern-like leaves. Inflorescences are either single, racemes, or panicles produced terminally or from leaf axils. The showy flowers, are dependent, heart-shaped and come in colors ranging from red, pink, white, purple, and yellow (Figure 142). The fruits are dehiscent or indehiscent capsules containing numerous seeds.

Toxic Principle and Mechanism of Action
A variety of mildly toxic isoquinoline alkaloids are present in all parts of the plant. Depending upon the species, apomorphine, cularine, protoberberine alkaloids may be present.[1] Either individually or in combination the alkaloids

have neurologic effects through their antagonistic action on the neurotransmitter gamma amino butyric acid (GABA.)[1,2] Similar isoquinoline alkaloids are found in the related genus *Corydalis* (fitweed, fumeroot), and *Fumaria* (fumitory).

Risk Assessment
Poisoning from *Dicentra* species is uncommon, but because these plants are commonly grown as garden plants, their potential toxicity should be recognized. Cattle and sheep are more commonly affected.

Clinical Signs
Muscle trembling and a staggering gait may be first noticed.[2] Excessive salivation and regurgitation of ingesta may also occur. Muscle tremors progress to the point animals are unable to stand, become recumbent, and develop tetanus-like seizures. These neurologic episodes may last 30 minutes at which time the animal appears normal. Fatalities are rare, as the animals are unable to eat sufficient quantities of the plant once neurologic signs begin.

References
1. Preininger V: Chemotaxonomy of Papaveraceae and Fumariaceae. In The Alkaloids, vol 29. Brossi A ed. Academic Press, Orlando, Fl pp 1-98, 1986.
2. Burrows GE, Tyrl RJ: Toxic Plants of North America. Iowa State University Press, Ames. pp 701-709, 2001.

Dieffenbachia
Family: Araceae

Common Name
Dumb cane, camilichigui, American arum, poison arum

Plant Description
Dieffenbachia species (25-30) originate in tropical regions of the Americas, and are universally used as ornamental plants in gardens and households. Dieffenbachias are one of the most popular houseplants in North America. The evergreen, erect perennials can grow up to 10 feet in height on thick stems that have prominent leaf scars. The leaves are simple, oblong-ovate, and are typically mottled or

Figure 143 *Dieffenbachia sequine*

variegated in various shades of white, cream, yellow, green, or red, the lighter colors occupying the space between the leaf veins (Figures 143 and 144). The petioles are long and sheath the stem. The inflorescence consists of a solitary spathe, whose margins overlap forming a tube. The fruits consist of red-yellow berries.

Figure 144 *Dieffenbachia sequine* 'Tropic Makianne'

The two most commonly encountered species of *Dieffenbachia*, and the ones from which most hybrids have been developed are *Dieffenbachia maculata*, and *D. seguine*

Toxic Principle and Mechanism of Action

Like other members of the Arum family, *Dieffenbachia* species contain calcium oxalate crystals in the stems and leaves.[1] The calcium oxalate crystals (raphides) are contained in specialized cells referred to as idioblasts.[1,2] Raphides are long needle-like crystals that are bunched together in these specialized cells. When the plant tissue is chewed by an animal, the crystals are extruded into the mouth and mucous membranes of the unfortunate animal. The raphides once embedded in the mucous membranes of the mouth cause an intense irritation and inflammation. In addition, there is evidence that the oxalate crystals act as a means for introducing other toxic compounds from the plant such as prostaglandins, histamine, and proteolytic enzymes that mediate the inflammatory response.[3] Unlike other members of the Arum family, Dieffenbachias are more toxic, because they have raphides in both the epidermal and mesophyll layers of the stems and leaves.[1]

The name "dumb cane" given to *Dieffenbachia* originated from the fact that people who chewed and ate the stems and leaves of the plant were unable to speak because they developed severe swelling of the mucous membranes of the mouth and pharynx. Similarly, dogs that chew on the plant may develop a severe stomatitis that has led to asphyxiation and death.[4]

Risk Assessment

Of all the Arum family, *Dieffenbachia* species are the most likely to cause problems in household pets, because the plants are common ornamentals in gardens and households. Dieffenbachia species make up one of the most frequent reports of plant poisoning exposures in people and animals reported to poison control centers.[5,6] Cattle and sheep may also be poisoned if they consume the plants.[7] In North America, this is most likely to occur when plant prunnings are inadvertently fed to livestock. Dieffenbachias should not be put into aviaries as birds are susceptible to poisoning from chewing on the leaves and stems.[8]

Clinical Signs

Dogs and cats that chew repeatedly on the leaves and stems of the *Dieffenbachia* develop edema of the oral mucous membranes shortly after chewing on the plants. Excessive salivation, difficulty in eating and swallowing, and vomiting are common signs of poisoning.[9,10] The edema, inflammation, and pain in the mouth of affected animals can resemble the lesions that would occur if the animal had consumed a caustic chemical. Humans who have chewed on Dieffenbachia stems and leaves also develop a severe stomatitis, leading to difficulty in eating and speaking.[11] If swelling in the pharynx is severe, animals will have difficulty in breathing, and in severe cases may die from asphyxiation.[9,12,13] Fortunately, in most cases, animals do not persist in chewing Dieffenbachia because of the rapid onset of inflammation and pain induced by the embedded oxalate crystals. Severe conjunctivitis and keratitis may result, if plant juices are rubbed into the eyes.[14]

Treatment
Unless salivation and vomiting are excessive, treatment is seldom necessary. Swelling of the lips and gums may persist for several days. Anti-inflammatory therapy may be necessary in cases where stomatitis is severe. The plant should be removed from the animal's environment.

References
1. Genua JM, Hillson CJ: The occurrence, type, and location of calcium oxalate crystals in the leaves of 14 species of Araceae. Ann Bot 56: 351-361, 1985.
2. Franceschi VR, Horner HT: Calcium oxalate crystals in plants. Bot Rev 46: 361-427, 1980.
3. Saha BP, Hussain M: A study of the irritating principle of aroids. Indian J Agric Sci 53: 833-836, 1983.
4. Loretti PL, da Silva MR, Ribiero RES: Accidental fatal poisoning of a dog by *Dieffenbachia picta* (dumb cane). Vet Hum Toxicol 45: 233-239, 2003.
5. Krenzelok EP, Jacobsen TD, Aronis JM: A review of 96,659 Dieffenbachia and philodendron exposures. J Toxicol Clin Toxol 34: 601, 1996.
6. Campbell A: Poisoning in small animals from commonly ingested plants. In Practice 20 :587-591, 1998.
7. Tokarnia CH et al: Experiments on the toxicity of some ornamental plants in cattle. Pesq Vet Bras. 16: 5-20, 1996.
8. Arai M, Stauber E, Shropshire CM: Evaluation of selected plants for their toxic effects in canaries. J Am Vet Med Assoc 200: 1329-1331, 1992.
9. Pedaci L, Kreneiok EP, Jacobsen TD et al: Dieffenbachia species exposure: an evidence-based assessment of symptom presentation. Vet Hum Toxicol 41: 335-338, 1999.
10. Knight MW, Dorman DC: Selected poisonous plant concerns in small animals. Vet Med 92: 260-272, 1997.
11. Gardner DG: Injury to the oral mucous membranes caused by the common houseplant, *Dieffenbachia*-a review. Oral Surg Oral Med Oral Pathol 631-633, 1994.
12. Wiese M, Kruszewska S, Kolacinski Z: Acute poisoning with *Dieffenbachia picta*. Vet Hum Toxicol 38: 356-358, 1996.
13. McGovern M: Botanical briefs: dumb cane-*Dieffenbachia picta* (lodd) Schott. Cutis 2000, 66: 333-334, 2000.
14. Seet B, Chan WK, Ang CL: Crystalline keratopathy from *Dieffenbachia* plant sap. Br J Ophthalmol 79: 78-99, 1995.

Family: Scrophulariaceae

Figure 145 *Digitalis purpurea*

Common Name
Foxglove, fairy bells, fairy gloves, lady's thimbles, digitalis

Plant Description
Digitalis species, of which there are 22, are native to Europe, northern Africa and western Asia. The most common species introduced into North America include *Digitalis lanata* (Grecian foxglove), *D. lutea* (straw foxglove), *D. purpurea* (common foxglove). The latter has become naturalized in the Pacific Northwest.

Perennial or biennial erect herbs, glabrous or tomentose, with flowering stems attaining heights of 3-6 feet (30-180 cm) depending upon the species. Leaves are alternate, lanceolate to ovate with entire or dentate margins. The flowers are produced on long terminal racemes, individual flowers being tubular with 5 fused sepals and five lobed petals. The white, pink, yellow, or brownish flowers, commonly contain spots or streaks in the throat of the flower (Figures 145-147). The fruits are conical capsules.

Toxic Principle and Mechanism of Action
The toxicity of digitalis species is attributed to numerous cardenolides, the best-known of which are digitoxin and digoxin. All parts of the plant are toxic, especially the seeds. The plant is also toxic when dried. The primary action of the digitalis cardenolides is on the cell membrane, where interference with normal transport of sodium and potassium ions across the

Figure 146 *Digitalis* cultivar

cell membrane occurs allowing an influx of intracellular calcium.[1] At low doses, myocardial function is improved, but at high doses cardiac conduction is impaired with resulting arrhythmias, heart block, and death.[2]

Other common garden plants containing similar cardenolides include butterfly weed *(Asclepiasa tuberosa)*, Lily of the valley *(Convallaria majalis)*,[3,4] oleander *(Nerium oleander)*, yellow oleander *(Thevetia thevetiodes)* and dogbane *(Apocynum species)*.[5]

Risk Assessment

Foxgloves are common garden plants, but rarely cause poisoning in household pets. Most cases of animal poisoning occur when livestock grazing on the plants, or they are given garden clippings containing the plants. Teas or herbal remedies made from *Digitalis* are a cause of human poisoning.[6] The water in vases containing the cut stems of foxgloves can contain sufficient dissolved cardenolides to be toxic.

Figure 147 *Digitalis* cultivar

Clinical Signs

Vomiting and diarrhea are common early sign of digitalis toxicity. This is followed by weakness, rapid heart rate, and changes in cardiac conduction with resulting decrease is in cardiac output, hypotension, collapse, and death. Early in the course of poisoning, the electrocardiogram may show an increasing P-R interval, sinus bradycardia, heart block, and ventricular ectopic beats. Hyperkalemia and hypocalcemia may develop. Induction of vomiting, gastric lavage, or administration of activated charcoal is appropriate for removing the plant and preventing further absorption of the toxins. Cathartics may also be used to help eliminate the plant rapidly from the digestive system. Serum potassium levels should be closely monitored and appropriate intravenous fluid therapy initiated as necessary. Phenytoin, as an antiarrhythmic drug effective against supraventricular and ventricular arrhythmias can be used as necessary.[7] The use of commercially available digitalis-specific antibody (Digibind – Burroughs Wellcome) may be a beneficial in counteracting the effects of the cardenolides.[7,8]

References

1. Ooi H, Colucci: Digitalis and allied cardiac glycosides. In Goodman and Gilman's The Pharmacological Basis of Therapeutics, 10th ed, hardman JG, Linbird LEP eds, McGraw-Hill, New York pp 916-921, 2001.
2. Detweiler DK, Knight DH: congestive heart failure in dogs: therapeutic concepts. J Am Vet Med Assoc 171: 106-114, 1977.
3. Kopp B, Kubelka W: New cardenolides of *Convallaria majalis*. Planta Med 45: 87-94, 1982a.
4. Kopp B, Kubelka W: New cardenolides of *Convallaria majalis*. Planta Med 45: 195-202, 1982b.
5. Singh B, Rastogi RP: Cardenolides-glycosides and genins. Phytochem 9: 315-331, 1970.
6. Dickstein ES, Kunkel FW: Foxglove tea poisonings. Am J Med 69: 167-169, 1980.
7. Gfeller RW, Messonier SP: Handbook of small animal toxicology and poisoning 2nd ed. Mosby, St. Louis, Missouri pp 161-162, 2004.
8. Smith TW et al: Treatment of life-threatening digitalis intoxication with digitoxin-specific Fab antibody fragments. Experience and 26 cases. N Eng J Med 307: 1357-1362, 1982.

Dracaena

Family: Agavaceae

Figure 148 *Dracaena* hybrid

Figure 149 *Dracaena deremensis*

Common Name
Corn plant, ribbon plant, dragon tree, money tree (Hawaii), lucky bamboo

Plant Description
A genus of some 40 species of evergreen shrubs or trees, originating from equatorial Africa and Asia, *Dracaena* species are commonly grown as foliage plants indoors or as ornamentals in tropical areas. Erect, slow-growing, branching in some species, shrubs or small trees growing to 20 ft. (6 m) in height. Leaves are simple, spirally arranged, with leathery, linear-lanceolate blades that often have white, ivory, or red margins (Figures 148-151). Leaf scars are usually prominent on the stems. Inflorescences are terminal panicles, with numerous white or yellowish white fragrant flowers. Fruits are yellow to red berries.

Some taxonomists have separated a number of the *Dracaena* into another genus *Pleomele.*

Toxic Principle and Mechanism of Action
A variety of steroidal saponins and glycosides have been identified from various species of *Dracaena.*[1] The toxicology of these compounds has not been defined, and is presumed to be related to the irritant effects of the saponins upon the gastrointestinal tract.

Risk Assessment
As common indoor plants, *Dracaena* species are frequently chewed and eaten by pets. Occasionally, the toxic effects encountered in cats necessitates the removal of the plants from the animal's environment.

Clinical Signs
Vomiting occasionally with blood, excessive salivation, anorexia, depression, ataxia and weakness may be shown by dogs and cats eating the plants. Dilated pupils, dyspnea, abdominal pain, and tachycardia may also be observed in cats.[2]

Treatment is seldom necessary once the animals are prevented from eating the plants. Symptomatic treatment, to prevent dehydration from vomiting may be necessary.

References

1. Gonzalez AG et al: Steroidal saponins from the bark of *Dracaena draco* and their cytotoxic activities. J Natural Products 66: 793-8, 2003.
2. Household Plant Reference. ASPCA National Animal Poison Control Center. 424 E. 92nd St., New York, NY 10128

Figure 150 *Dracaena marginata*

Figure 151 *Dracaena sanderiana* (Lucky Bamboo)

Epipremnum

Family: Araceae

Common Name

Pothos, golden pothos, hunter's robe, ivy arum, devil's ivy

Plant Description

Comprising a genus of 8 species, these evergreen vine-like plants are native to the tropical areas of the Pacific and Southeast Asia. Specialized adhesive aerial roots enable the plants to climb to great heights (60 feet) in trees when grown in their natural habitat. The leaves are

Figure 152 *Epipremnum pinnatum*

Figure 153 *Epipremnum pinnatum aureum*

Figure 154 *Epipremnum pinnatum* leaves

simple, ovate to lanceolate, with juvenile and adult forms, and are either green or variegated. The leaves have long petioles, sheathing the stem at the base, and have prominent parallel-convergent veins. The inflorescence consists of a solitary green, yellow, or purple spathe, with the spadix shorter than the spathe. The edges of the spathe do not overlap to form tubes. Rarely do the plants flower as indoor plants unless they have reached considerable size, which is unlikely to occur unless they are grown outdoors (Figures 152-154).

The 2 most commonly cultivated species of pothos are:
Epipremnum pinnatum (Raphidophothora pinnata) - Golden pothos is a common variety correctly named *E. pinnatum* 'Aureum'

E. pictum (Scindapsus pictus) - silver vine. This species is smaller and has silver colored leaf markings.

The genera *Epipremnum, Syngonuim* and *Scindapsus* closely resemble each other (See Syngonuim).

Toxic Principle and Mechanism of Action
Like other members of the Araceae family, *Epipremnum* species contain oxalate crystals in the stems and leaves.[1,2] The calcium oxalate crystals (raphides) are contained in specialized cells referred to as idioblasts.[1,2] Raphides are long needle-like crystals bunched together in these specialized cells, and when the plant tissue is chewed by an animal, the crystals are extruded into the mouth and mucous membranes of the unfortunate animal. The raphides once embedded in the mucous membranes of the mouth cause an intense irritation and inflammation. Evidence exists to suggest that the oxalate crystals act as a means for introducing other toxic compounds from the plant such as prostaglandins, histamine, and proteolytic enzymes that mediate the inflammatory response[3] (See Dieffenbachia).

Risk Assessment
Epipremnum species are frequently grown for their striking foliage and consequently, household pets have access to the plants.

Clinical Signs

Dogs and cats that chew repeatedly on the leaves and stems of Epipremnums may salivate excessively and vomit as a result of the irritant effects of the calcium oxalate crystals embedded in their oral mucous membranes. The painful swelling in the mouth may prevent the animal from eating for several days. Severe conjunctivitis may result if plant juices are rubbed in the eye.

Treatment

Unless salivation and vomiting are excessive, treatment is seldom necessary. Anti-inflammatory therapy may be necessary in cases where stomatitis is severe. The plants should be removed or made inaccessible to the animals eating them.

References

1. Genua JM, Hillson CJ: The occurrence, type, and location of calcium oxalate crystals in the leaves of 14 species of Araceae. Ann Bot 56: 351-361, 1985.
2. Franceschi VR, Horner HT: Calcium oxalate crystals in plants. Bot Rev 46: 361-427, 1980.
3. Saha BP, Hussain M: A study of the irritating principle of aroids. Indian J Agric Sci 1983, 53: 833-836.

Eriobotrya

Family: Rosaceae

Common Name

Loquat, Japanese plum, Japanese medlar, - *Eriobotrya japonica.*

Plant Description

A genus of about 30 species of evergreen shrubs or trees native to temperate China, Japan, and many parts of Asia. Trees may grow to 9 m. (30ft), and have large, prominently veined leathery leaves that have a white hairy, velvet-like underside (Figure 155). Sprays of white

Figure 155 *Eriobotrya japonica*

scented flowers are produced at the ends of branches, and yield fruits which turn yellow when ripe.[1] The flesh of the fruits is edible, but the single or paired brown seeds in the fruits are not edible.

Toxic Principle and Mechanism of Action

The cyanogenic glycoside amygdalin is present in the seeds of the loquat.[2] Ingestion of well chewed seeds can result in cyanide poisoning in animals.[3] Amygdalin, once hydrolyzed in the stomach, produces hydrogen cyanide (Prussic acid) which is readily absorbed into the blood where it blocks the cytochrome oxidase in the red blood cells. This results in acute cellular anoxia and death of the animal. A variety of terpenoid compounds are also present in the plant but their toxicologic significance has not been established.[4]

Risk Assessment
Poisoning is unlikely unless the seeds of the fruits are eaten by animals or humans.

Clinical Signs
Once the cyanogenic glycosides are hydrolyzed in the stomach to release hydrogen cyanide, acute deaths may result. Animals can be expected to exhibit sudden onset of respiratory difficulty, cyanosis, and death.

Treatment
Affected animals should be given a solution of sodium thiosulfate orally and intravenously when in acute respiratory distress. Care must be taken to avoid stressing the animal, and treatment must be initiated early in the course of poisoning if treatment is to be successful. Other supportive therapy including oxygen and intravenous fluids may be necessary.

References
1. Spoerke DG, Spoerke SE, Rumack BH: Berry identification using a modified botanic key. Vet Human Toxicol, 30, 260-264, 1988.
2. Seigler DS: The naturally occurring cyanogenic glycosides. Reinhold L et al (eds) Progress in phytochemistry. 4: 83-120, 1977.
3. Weber MA. Garner M. Cyanide toxicosis in Asian small-clawed otters *(Amblonyx cinereus)* secondary to ingestion of loquat *(Eriobotrya japonica)*. J Zoo & Wildlife Medicine. 33: 145-6, 2002.
4. Lee TH, Lee SS, KUO WC, Chou CH: Monoterpene glycosides and triterpene acids *Eriobotrya* japonica. J Nat Products 64: 865-869, 2001.

Erythrina

Family: Fabaceae

Figure 156 *Erythrina crista-galli*

Common Name
Coral tree, coral bean

Plant Description
There are over 100 species of *Erythrina*, native to the tropical areas of the Americas and Africa. The evergreen or deciduous, perennial shrubs or trees typically have blunt conical thorns on the trunks and branches. The leaves are pinnately compound with 3 ovate to rhomboid leaflets. Inflorescences are large showy, bright red to orange-yellow, terminal or axillary racemes (Figures 156 and 157). Each flower has 5 fused sepals, and the 5 petals form a distinctive pea-like banner and keel enclosing the stamens. Depending on the species, many

bright red, or brown seeds, some with black spotting are produced in leguminous pods.

Toxic Principle and Mechanism of Action

A variety of unique, complex alkaloids are found in the various species of *Erythrina*.[1,2] The alkaloids are present in all parts of the plant, but especially in the flowers and seeds. All the alkaloids have toxic effects when ingested and have primarily a curare like effect, causing paralysis. The alkaloids are passed through the milk of animals that chew and eat the seeds.[3]

Risk Assessment

Figure 157 Erythrina herbacea

Since numerous species of *Erythrina* have showy flowers they are frequently cultivated as ornamental shrubs or trees in gardens, parks, and along streets in tropical areas. The colorful red seeds are a potential hazard to children and animals that might chew and swallow them. Although there are no documented cases of *Erythrina* poisoning in household pets, the seeds have been reportedly used to poison dogs in Mexico.[4]

Clinical Signs

Animals that have consumed the seeds of flowers of *Erythrina* can be anticipated to develop a curare-like paralysis that would manifest clinically as muscle weakness and paralysis. Treatment should be symptomatic.

References

1. Chawla AS, Kapoor VK: Erythrina alkaloids. In alkaloids: chemical and biological perspectives,vol 9. Pelletier SW ed, Elsevier, New York 1995, pp 85-153, 1995.
2. Burrows GE, Tyrl RJ: Toxic Plants of North America. Iowa State University Press, Ames pp 552-555, 2001.
3. Soto-Hernadez M, Jackson AH: studies of alkaloids in foliage of *Erythrina berteroana*, and *Erythrina poeppigiana:* detection of beta-erythroidine in goats milk. Phytochem Anal 4: 97-99, 1993.
4. Morton JF: Plants Poisonous to People in Florida. 2nd ed. JF Morton ed. Southeastern Printing, Stuart, Florida pp 149, 1999.

Eucharis

Family: Liliaceae

Figure 158 *Eucharis amazonica*

Common Name
Amazon lily, eucharist lily

Plant Description
Native to South America, *Eucharis* are bulbous perennials that prefer warm, shady growing conditions. In temperate zones they may be grown as potted house plants. The large showy white marked with green flowers, are fragrant and are produced in groups of 4 at the tops of tall flower stalks (Figure 158).

Toxic Principle and Mechanism of Action
At least 15 phenanthridine alkaloids including lycorine, have been identified in the leaves, stems, and bulbs of Eucharis.[1,2] The concentrations of the alkaloids are highest in the outer layers of the bulbs. The total alkaloid concentration in the leaves and parts of the bulbs is reported as 0.5%. Phenanthridine alkaloids have been isolated from other genera of the Amaryllis family, including species of *Amaryllis, Clivia, Galanthus, Hippeastrum, Haemanthus, Hymenocallis, Leucojum, Narcissus, Nerine, Sprekelia, and Zephranthes.*

Risk Assessment
The greatest risk is to household pets that eat the bulbs of the plant.

Clinical Signs
Vomiting, excessive salivation, abdominal pain, diarrhea, and difficulty in breathing, are associated with the phenanthridine alkaloids present in the lily family. If large quantities of the leaves, stems, and bulb are consumed, depression, ataxia, seizures, bradycardia, and hypotension may develop. Poisoning is rarely fatal, and can generally be treated symptomatically.

References
1. Martin SF: The Amaryllidaceae alkaloids. In The Alkaloids: Chemistry and Physiology. vol 30, Brossi A (ed) Academic Press, San Diego, Calif 251-376, 1987.
2. Cabezas F, Ramirez A, Viladomat F, Codina C, Bastida J: Alkaloids from *Eucharis amazonica* (Amaryllidaceae).Chem Pharmaceut Bulletin 51: 315-317, 2003.

Euonymus

Family: Celastraceae

Common Name

A variety of common names are attributed to *Euonymus* depending upon the species. The most commonly encountered species include winged euonymus, spindle tree or burning bush *(E. alatus)*, arrow wood, wahoo *(E, atropurpureus)*, European spindle tree *(E. europaeus)*, evergreen euonymus *(E. japonicus)*.

Plant Description

A cosmopolitan genus of approximately 200 species, *Euonymus* are woody shrubs or small trees that may be evergreen or deciduous, with stems that are either round or angular or winged. Leaves are petioled, elliptic to ovate, with serrate margins. Flowers are produced singly or as cymes from leaf axils, and are small with 4-5 fused sepals and 4-5 petals greenish to purple in color. Fruits are 3-5 lobed yellow – brown capsules that split open to reveal seeds that have orange to red arils. Fruits are similar in appearance to those of bittersweet *(Celastrus scandens)* (Figures 159-162).

Figure 159 *Euonymus elatus*

Figure 160 *Euonymus japonicus* fruits

Toxic Principle and Mechanism of Action

A variety of toxic alkaloids and cardenolides have been isolated from *Euonymus* species.[1-3] The seeds and to a lesser extent the leaves contain the cardenolides, while the alkaloids are found in all parts of the plants.[4] The cardenolides evonoside, evomonoside and others have a digitalis-like effect on the heart, while the effects of the alkaloids are poorly understood. Triterpenoids have also been isolated from *E. europaeus*.[5] Similar cardenolides are found in bittersweet *(Celastrus scandens)*, but cases of poisoning have not been recorded.

Risk Assessment

Poisoning from *Euonymus* species is rarely reported even though the plants are commonly grown for their attractive foliage, especially in the Fall when the leaves in some species turn bright red. The berries are also attractive and persist on the branches after the leaves have fallen.

Figure 161 *Euonymus europaeus*

Figure 162 *Euonymus oxyphylus* fruits

Clinical Signs

Diarrhea, abdominal pain, constipation, vomiting, and weakness are the most frequently reported signs of poisoning. Cardiac dysrhythmias may be encountered in severe cases.[6]

Treatment is seldom necessary and is generally symptomatic and directed towards relieving abdominal pain, diarrhea and cardiac irregularities.

References

1. Burrows GE, Tyrl RJ. Toxic Plants of North America. Iowa State University Press, Ames. pp 336-340, 2001.
2. Baek NI, Lee YH, Park JD, Kim SI, Ahn BZ. Euonymoside A: a new cytoxic cardenolides glycoside from the bark of *Euonymus sieboldianus*. Planta Med 60: 26-28, 1994.
3. Kitanaka S. Takido M. Mizoue K. Nakaike S. Cytotoxic cardenolides from woods of *Euonymus alata*. Chem Pharmaceutical Bulletin. 1996, 44: 615-617.
4. Ishiwata H, Shizuri Y, Yamada K. Three sesquiterpene alkaloids from *Euonymus alatus* forma striatus. Phytochemistry 1983, 22: 2839-2341.
5. Pasich B, Bishay DW, Kowalewski Z, Rompel H. Chemical investigation of *Euonymus europaeus*. Planta Med 1980, 39: 391-395.
6. Lampe KF, McCann MA. AMA Handbook of Poisonous and Injurious Plants. Am Med Assoc, Chicago, Illinois 1985, pp 79-81.

Eupatorium rugosum (Ageratina altissima)

Family: Asteraceae

Common Name
White snakeroot, rich weed, hemp agrimony

Plant Description
A perennial woodlands plant of the eastern, south eastern Great Plains areas of North America, growing to 1.5m (5 ft.) in height, spreading by rhizomatous roots, and with erect stems from a woody base. Leaves are opposite, simple, ovate (6-15 cm long, 3-12 cm wide), upper leaves smaller, petiolate and with serrated edges. The inflorescence are flat topped or domed heads with 12-24 white disk florets (Figures 163 and 164). The fruits are black achenes (2-3 mm long), 5-angled, with hairy pappus.

Figure 163 *Eupatorium rugosum*

The genus *Eupatorium* consists of about 40 species native to North America, with a few species from South America and Asia. Closely related to *Eupatorium* is the genus *Ageratina*, some taxonomists considering them the same genus.
Only two species of *Eupatorium* and *Ageratina* are known to be toxic and these are:
Eupatorium rugosum (Ageratina altissima) – white snakeroot
Ageratina adenophora (E. adenophora) – Crofton weed.
Another genus belonging to the tribe Eupatorieae, and very similar to the above genera is the genus *Ageratum*. It is commonly represented in cultivation by *A. houstonianum*, and is reported to be toxic.

Toxic Principle and Mechanism of Action
White snakeroot contains a complex mixture of sterols and ketones (benzofurans) collectively referred to as Tremetol, a name that is descriptive of the muscle tremors seen clinically.[1,2] The primary action of tremetol appears to be associated with enzyme inhibition in the tricarboxylic acid cycle, with resulting depletion of liver glycogen, hypoglycemia, and cellular damage in the form of hepatocellular necrosis and myocardial degeneration.[3] Tremetol is excreted in the milk of lactating animals and therefore poses a risk to the suckling animal, or to humans drinking the milk.[4] "Milk sickness" is the term given to a disease syndrome of weakness, muscle tremors, and death that affected early settlers in the eastern United States. Whole populations of cattle and many people who drank the milk from cows that had been grazing on white snake root died from "milk sickness" until it was discovered that the cause of the problem was the white snakeroot. Abraham Lincoln's mother reputedly died from "milk sickness" in 1818.[5] Tremetol is cumulative in effect, its concentration being highest in the green plant, with the dried plant being considerably less toxic. The lethal dose of green plant in goats has been determined to be 5mg/kg body weight.[6]

Meat from animals that have died from white snakeroot poisoning is suspected of causing poisoning in carnivores. Pasteurization of milk does not destroy tremetol, and the fact that it is fat soluble means that butter can contain tremetol.[7]

Risk Assessment
White snakeroot is included not because it is likely to be a problem to household pets, but because it is common in woodland areas of much of eastern North America, and has historical toxicologic significance. It is also important because the milk from lactating cows or goats that have been grazing on white snakeroot

Figure 164 *Eupatorium rugosum*

will contain the toxic tremetol that can induce tremors in animals and people who drink the milk.

Clinical Signs
There is considerable variation in the clinical signs shown by different species of animal. People with "milk sickness" develop weakness, muscle tremors, nausea, and death. Horses develop excessive sweating, difficulty in swallowing, and cardiac abnormalities.[8] Goats develop neurologic signs associated with severe hepatic necrosis causing an encephalopathy.[6]

There is no specific treatment for tremetol poisoning, and therapy should be directed at providing glucose, intravenous electrolytes, and other supportive treatment as necessary.

References
1. Beier RC, et al. Isolation of the major components in white snakeroot that is toxic after microsomal activation: possible explanation of sporadic toxicity of white snake root plants and extracts. Natural Toxins, 286-293, 1993.
2. Sharma OP, Dawra RK, Kurade NP, Sharma PD. A review of the toxicosis and biological properties of the genus *Eupatorium*. Review. Natural Toxins. 1998, 6: 1-14, 1998.
3. Beier RC, Norman JO. the toxic factor in white snake root: identity, analysis and prevention. Vet Hum Toxicol 32 Suppl: 81-88, 1990.
4. Panter KE, James LF. Natural toxicants in the milk: a review. J Anim Sci 68: 892-904, 1990.
5. Duffy DC. A land of milk and poison. Nat Hist 7: 4-8, 1990.
6. Reagor JC, Jones LP, Ray AC, Baily EM. *Eupatorium rugosum*, white snakeroot poisoning in goats. Third International Symposium on Poisonous Plants. Logan Utah 1989
7. Burrows GE, Tyrl RJ. Toxic Plants of North America. Iowa State University Press, Ames. pp 161-165,. 2001.
8. Thompson LJ. Depression and choke in a horse: probable white snakeroot toxicosis. Vet Hum Toxicol 31: 321-322, 1989.

Euphorbia

Family: Euphorbiaceae

Common Name

There are a large number of different names given to the many spurges or euphorbias that are commonly grown as garden or house plants. The more common ones include: Poinsettia *(E. pulcherrima)*, crown of thorns *(E. milii)*, snow on the mountain (E. marginata), pencil cactus *(E. tirucalli)*, creeping spurge *(E. myrsinites)*, hat rack cactus *(E. lactea)*, candelabra cactus *(E. candelabrum)*

Figure 165 *Euphorbia pulcherrima*
Inset-actual flower or cyanthium

Plant Description

Some 2000 species of *Euphorbia* are native to a wide variety of climates across the world. This diverse genus ranges from small prostrate herbs to shrubs, and trees to 10-15m in height. Some are covered with spines and are succulent and cactus-like. All have a viscid milky sap. Stems are prostrate or erect, succulent or not. Leaves are simple, alternate or opposite, petiolate or sessile, and in some species very small and deciduous. Characteristic of all *Euphorbia* species is the inflorescence called a cyathium (Figure 165). Resembling a single flower, it is actually a single pistillate flower surrounded by many staminate flowers, all of which are enclosed in an urn-like structure. Sepals and petals are absent. Many species have showy bracts surrounding the cyathia, best exemplified by the poinsettia *(E. pulcherrima)* (Figure 166). Fruits are 3-lobed capsules containing 3 seeds Figures 167-170 illustrate the variety of species.

Figure 166 *Euphorbia pulcherrima* cultivar

Toxic Principle and Mechanism of Action

Depending upon the species, a variety of diterpenoid euphorbol esters are found in

Figure 167 *Euphorbia milii*

all parts of the plants. These compounds are irritants and can cause dermatitis in some individuals handling the plants, and can cause corneal ulcers if introduced

Figure 168 *Euphorbia tirucalli*

Figure 169 *Euphorbia myrsinites*

into the eyes.[1-4] Different *Euphorbia* species have been used as fish poisons in Africa.[5] There is marked variation in the variety and composition of diterpenoids present in the different species, one of the more toxic species being *E. tirucalli*, The most popular of all houseplants, *E. pulcherrima* (poinsettia), especially its numerous hybrids, is not toxic unless consumed in considerable quantities, far more than would normally be available in a large poinsettia plant![6,7]

Euphorbia peplus, has been shown to be toxic to the liver and kidney endothelial and parenchymal cells causing death in goats. The toxic substances in the plant are also passed through the milk and can cause similar lesions in the kids drinking the milk.[8]

Risk Assessment
Euphorbias or spurges are one of the most frequently reported causes of plant poisoning at poison control centers.[9] However, the poinsettia, despite its reputation as a toxic plant, is not poisonous. A few hypersensitive individuals may develop a skin rash if they handle the plants excessively. Conjunctival irritation and corneal ulcers may occur if the milky sap is introduced into the eyes. Cats that chew on the plants may salivate and vomit. Symptoms of poisoning are usually of short duration and rarely need treatment. Those species of *Euphorbia* that have spines *(E. lactea)* can cause mechanical injury.

Clinical Signs
Mouth irritation, salivation, and vomiting can be anticipated in cats and dogs that might chew or eat any of the *Euphorbia* species. The milky sap if rubbed into the eyes can cause conjunctivitis, lacrimation, and in the worst cases corneal ulcers.[1,3] Contact with the milky sap can result in dermatitis especially in people who handle the plants excessively.[4]

Most exposures to *Euphorbia* species do not require specific treatment. Removal of the resinous milky sap from the hair coat of animals or the skin of people is best accomplished using mild soap or alcohol. Anti-inflammatory drugs may be helpful in cases where dermatitis is severe. Persisting eye irritation should be treated by an ophthalmologist.

References

1. Hsueh KF, Lin PY, Lee SM, Hsieh CF: Ocular injuries from plant sap of genera *Euphorbia* and *Dieffenbachia.* J Chinese Med Assoc 67: 93-98, 2004.

2. Paulsen E, Skov PS, Andersen KE: Immediate skin and mucosal symptoms from pot plants and vegetables in gardeners and greenhouse workers. Contact Dermatitis. 39: 166-170, 1998.

3. Scott IU, Karp CL: Euphorbia sap keratopathy: four cases and a possible pathogenic mechanism. British J Ophthalmology. 80: 823-826, 1996.

4. Worobec SM, Hickey TA, Kinghorn AD, Soejarto DD, West D: Irritant contact dermatitis from an ornamental Euphorbia. Contact Dermatitis 7: 19-22, 1981.

Figure 170 *Euphorbia lactea*

5. Neuwinger HD. Plants used for poison fishing in tropical Africa. Toxicon. 44: 417-430, 2004.

6. Krenzelok EP, Jacobsen TD, Aronis JM: Poinsettia exposures have good outcomes...just as we thought. Am J Emergency Med 14: 671-674, 1996.

7. Runyon R: Toxicity of fresh poinsettia *(Euphorbia pulcherrima)* to Sprague-Dawley rats. Clin Toxicol 16: 167-173, 1980.

8. Nawito M, Ahmed YF, Zayed SM, Hecker E: Dietary cancer risk from conditional cancerogens in produce of livestock fed on species of spurge (Euphorbiaceae). II. Pathophysiological investigations in lactating goats fed on the skin irritant herb *Euphorbia peplus* and in their milk-raised kids. J Cancer Res Clin Oncology 124: 179-185, 1998.

9. Krenzelok EP, Jacobsen TD: Plant exposures ... a national profile of the most common plant genera. Vet Hum Toxicol. 39: 248-249, 1997.

Ficus

Family: Moraceae

Figure 171 *Ficus elastica*

Figure 172 *Ficus elastica* 'Decora'

Common Name
Fig, ficus
The most commonly encountered species of fig include the common edible fig *(Ficus carica)*, rubber plant or Indian rubber tree *(F. elastica)*, fiddle leaf fig *(F. lyrata)*, banyan tree *(F. macrophylla)*, and weeping fig *(F. benjamina)*.

Plant Description
A genus of some 800 species of evergreen, branching, woody stemmed shrubs, trees or climbers, *Ficus* are native to most tropical areas of the world. Some species of *Ficus* strangle the host tree with their roots, eventually killing the tree and becoming massive trees themselves. Leaves are large, leathery, in some variegated, alternate, simple or deeply lobed (Figures 171-173). The leaves and stems of all *Ficus* contain a milky sap. Minute petalless flowers are produced on the swollen ends of short stalks produced in the leaf axils. The swollen stems enlarge around the flowers forming the fruit or fig (Figure 174).

Toxic Principle and Mechanism of Action
Depending on the species of *Ficus*, the sap contains a proteolytic enzyme ficin, and ficusin which is a phototoxic psoralen.[1] Fig dermatitis is encountered in some individuals who get the sap on their skin. Redness and blistering around the mouth is reported in some people who eat the figs. Exposure to sunlight will exacerbate the dermatitis.

Risk Assessment
Ficus species are commonly grown as garden plants in sub tropical and tropical areas for their attractive leaves and for their fruits in the case of the edible fig *Ficus carica*. In temperate areas *Ficus* species are frequently grown as potted indoor plants. Animal poisoning is unlikely, but people can develop a photodermatitis from getting the milky sap on their skin and then being exposed to ultra violet wave length light.

Clinical Signs

Dermatitis, especially after exposure to sunlight, is seen in some individuals. Blistering and inflammation around the mouth after eating the fruits of the fig can be severe in some hypersensitive people. These individuals develop a contact dermatitis not unlike that seen with poison ivy or oak. Long lasting purple discoloration of the skin is reported in some individuals contacting the milky sap.

Treatment may be required in severe cases of photodermatitis and a physician should be contacted. Washing the affected areas with mild soap, staying out of the sun, and if necessary anti-inflammatory drugs may be necessary.

References

1. Turner NJ, Szczawinski AF. Common Poisonous Plants and Mushrooms. Timber Press, Portland, Oregon pp 230-231, 1995.

Figure 173 *Ficus benjamina*

Figure 174 *Ficus carica*

Galanthus
Family: Amaryllidaceae

Common Name
Snow drop

Plant Description
A genus of about 19 species of small bulbous plants, *Galanthus* are native to Europe and Asia. Small white, nodding flowers are produced just before and above the narrow, linear leaves. The 3 inner petals are much shorter than the 3 outer ones, and usually have green

Figure 175 *Galanthus caucasicus*

Figure 176 *Galanthus caucasicus* flowers

markings. *Galanthus* are often one of the first bulbs to bloom in the spring (Figures 175 and 176).

Toxic Principle and Mechanism of Action

At least 15 phenanthridine alkaloids including lycorine, have been identified in the leaves, stems, and bulbs of *Galanthus*.[1] The concentrations of the alkaloids are highest in the outer layers of the bulbs. The total alkaloid concentration in the leaves and parts of the bulbs is reported as 0.5%. Phenanthridine alkaloids have been isolated from other genera of the Amaryllis family, including species of *Amaryllis, Clivia, Eucharis, Hippeastrum, Haemanthus, Hymenocallis, Leucojum, Narcissus, Nerine, Sprekelia, and Zephranthes*

Risk Assessment

Bulbs left accessible to household pets pose the greatest risk.

Clinical Signs

Vomiting, excessive salivation, abdominal pain, diarrhea, and difficulty in breathing, are associated with the phenanthridine alkaloids present in the lily family. If large quantities of the leaves, stems, and bulb are consumed, depression, ataxia, seizures, bradycardia, and hypotension may develop. Poisoning is rarely fatal, and can generally be treated symptomatically.

References

1. Martin SF: The Amaryllidaceae alkaloids. In The Alkaloids: Chemistry and Physiology. vol 30, Brossi A (ed) Academic Press, San Diego, Calif, pp 251-376, 1987.

Gelsemium

Family: Loganiaceae

Common Name

Carolina jessamine, yellow Jessamine, evening trumpet vine - *Gelsemium sempervirens*
Evening trumpet flower – *Gelsemium rankinii*

Plant Description

Consisting of two North American and one Asian species of *Gelsemium*, these evergreen, twining or trailing, perennials, have lanceolate to elliptic, shiny, green leaves with solitary, or 2-5 flowered cymes. The flowers are showy fragrant, trumpet-shaped, bright yellow with 5 sepals and 5 fused petals (Figures 177 and 178). Fruits are ovoid capsules with a short beak. The seeds are brown, flattened, and generally winged.

Figure 177 *Gelsemium sempervirens*

Toxic Principle and Mechanism of Action

Gelsemium species contain the neurotoxic alkaloids gelsemine, gelsemicine, gelsedine, gelseverine, and gelseminine.[1] The indole, sempervirine, is also considered a significant toxic component.

Figure 178 *Gelsemium sempervirens* flowers

The alkaloids are found in all parts of the plant, especially in the roots and act on nerve endings causing paralysis, muscle weakness, and clonic convulsions. At high doses the alkaloids act centrally on the nervous system against gamma amino butyric acid (GABA) causing convulsions and respiratory failure.[2]

Risk Assessment

Carolina jessamine is commonly grown as a garden plant for the profusion of yellow flowers it produces in the summer. Cases of human poisoning are reported and in Asia the plant has been used for suicidal purposes.[2,3] The risk of household pets being poisoned by eating the plant is minimal. Most cases of poisoning occur in livestock including geese that may graze on plant trimmings or the plant growing in or near their enclosure.[4]

Clinical Signs

In animals acutely poisoned by *Gelsemium* species, neurologic signs predominate, and are characterized by progressive weakness, convulsions, respiratory failure, and death.[4] Postmortem lesions are nonspecific. Histologically evidence of mild diffuse neuronal and cerebellar Purkinje cell loss with vacuolation of the brainstem and cerebral white matter may be present.[4]

References

1. Schun Y, Cordell GA: Cytotoxic steroids of *Gelsemium sempervirens*.
2. Rujjanawate C. Kanjanapothi D. Panthong A: Pharmacological effect and toxicity of alkaloids from *Gelsemium elegans* Benth. J Ethnopharm 89: 91-95, 2003.
3. Blaw ME, Adkisson MA, Levin D, Garriott JC, Tindall RS: Poisoning with Carolina jessamine (*Gelsemium sempervirens* [L.] Ait.). J Pediatrics 94: 998-1001, 1979.
4. Thompson LJ, Frazier K, Stiver S, Styer E: Multiple animal intoxications associated with Carolina jessamine *(Gelsemium sempervirens)* ingestions. Vet Human Toxicol 44: 272-272, 2002.

Gloriosa superba

Family: Liliaceae

Figure 179 *Gloriosa superba* 'Rothschildiana'

Common Name

Gloriosa lily, glory lily, flame lily, climbing lily, superb lily, tiger's claws

Plant Description

Consisting of one generally accepted species native to tropical Africa and Asia, this climbing vine-like lily with tendrils at the leaf tips, emerges from an underground irregular tuber. Several twining slender stems are produced with alternate or opposite glossy, lanceolate leaves with a terminal tendril. The solitary flowers are produced in the leaf axils on long pedicels. The showy flowers have similar red and yellow, wavy-edged, sepals and petals that are strongly reflexed, exposing the conspicuous stamens (Figure 179). The fruits consist of capsules with multiple seeds.

Toxic Principle and Mechanism of Action

Like the *Colchicum* species, *Gloriosa superba* contains up to 0.36% of colchicine in its tubers.[1] This alkaloid blocks cell mitosis causing multiple organ failure (See Colchicum). Consumption of the tuber by humans in Sri Lanka was the cause of 8 suicidal deaths, and *Gloriosa superba* was the most common cause of plant induced poisoning.[2,3] In parts of east and southern Africa, it is well recognized as toxic to both humans and animals.[4]

Risk Assessment

Although no reports of animal poisoning have been associated with *Gloriosa superba*, the tubers of this plant contain sufficient quantities of colchicine to make it one of the more toxic tubers that become accessible to household pets. Dogs have been reportedly killed by feeding them the ground tuber.[5]

Clinical Signs

Should the tuber be ingested the most likely effect would be severe vomiting. Diarrhea sweating, seizures, and cardiac irregularities may also develop.[2,3] Treatment should be aimed at managing the vomiting and diarrhea.

References

1. Nagaratnam N, DeSilva DPKM, DeSilva N: Colchicine poisoning following ingestion of *Gloriosa superba* tubers. Trop Geogr Med 25: 15-17, 1973.
2. Mendis S: Colchicine cardiotoxicity following ingestion of *Gloriosa superba* tubers. Postgraduate Med J 65: 752-755, 1989.
3. Fernando R. Fernando DN: Poisoning with plants and mushrooms in Sri Lanka: a retrospective hospital based study. Vet Human Toxicol 32: 579-581, 1990.
4. Watt J.M. and M.G. Breyer-Brandwijk. The Medicinal and Poisonous Plants of Southern and Eastern Africa. 2nd Ed. E. & S. Livingston, Edinburgh 700-707, 1962.
5. Morton Julia F. In Plants Poisonous to People in Florida. 2nd ed. South Eastern Printing Co. Stuart, Florida. pp 37-38, 1982.

Gymnocladus dioica

Family: Fabaceae

Common Name

Kentucky coffee tree

Plant Description

Native to North America and East Asia, the 4 *Gymnocladus* species are slow growing trees attaining heights of 70 feet. The Kentucky coffee tree *(Gymnocladus dioica)*, the North American representative of the genus, is indigenous to eastern North America, but has been planted in many other areas as an ornamental tree.

Deciduous, branching trees with large (1m in length), twice pinnately compound leaves, with pairs of ovate leaflets, 4-7 per

Figure 180 *Gymnocladus dioica*

pinna. Terminal leaflets are absent. The inflorescence is a terminal raceme or panicle. Flowers are small, star-shaped, whitish,and fragrant; the male and female flowers being produced on separate plants. The female trees produce 4-8 inch long, brown leguminous pods with 5-8 hard-coated, olive-brown seeds (Figures 180 and 181).

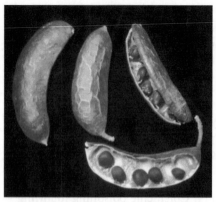

Figure 181 *Gymnocladus dioica* seed pods

Toxic Principle and Mechanism of Action

A water soluble, heat labile toxin or group of toxins are present in the leaves and seeds. A group of complex glycosides called gymnocladosapponins may be responsible for the toxicity associated with consumption of the uncooked seeds or leaves. The cooked seeds were at one time tried as a coffee substitute. Poisoning of livestock has been associated with animals drinking water into which the *Gymnocadus* seeds had fallen. The toxins appear to have gastrointestinal irritation and narcotic effects.[1]

Risk Assessment

Kentucky coffee trees are popular landscaping trees because of their shape and attractive foliage. The female trees in some years produce large numbers of pods which fall to the ground and are attractive to children, and dogs may chew on the pods. There are no documented cases of poisoning in dogs from eating the pods and seeds but the potential for poisoning is present.

Clinical Signs

Gastrointestinal irritation characterized by excessive salivation, colic, and diarrhea can be anticipated in animals eating the leaves or uncooked beans. Affected animals may show marked depression associated with the narcotic effects of the toxins.[2] Treatment when necessary should be directed at relief of colic and diarrhea.

References

1. Burrows GE, Tyrl RJ: Toxic Plants of North America. Iowa State University Press, Ames pp 558-560, 2002.
2. Kingsbury JM: Poisonous Plants of the United States and Canada. Prentice-Hall Inc. Englewood Cliffs New Jersey pp 322-323, 1964.

Haemanthus
Family: Liliaceae (Amaryllidaceae)

Common Name
Blood lily, powder puff lily, paint brush lily

Plant Description
A genus of approximately 20 species from southern and eastern Africa and the Arabian penninsula, *Haemanthus* lilies are deciduous or evergreen bulbous plants with striking large red inflorescences. The flowers are produced on naked stems and are followed by the leaves.

Figure 182 *Haemanthus multiflorus*

Many *Haemanthus* species are now reclassified by some taxonomists under the genus *Scadoxus*. (Figure 182).

Toxic Principle and Mechanism of Action
Although the specific toxins have not been identified, *Haemanthus* species contain phenanthridine alkaloids including lycorine, coccinine, montanine, hippeastrine, and haemanthidine.[1] The phenanthridine alkaloids are present in many of the Liliaceae, most notably in the *Narcissus* group. The bulbs appear to be the most toxic.[2]

Risk Assessment
Haemanthus have not been reported as a problem for household pets. However, with the ever increasing introduction of attractive and exotic plants into the horticultural market, there is the potential for the toxic bulbs to be accessible to household pets.

Clinical Signs
Vomiting, excessive salivation, abdominal pain, diarrhea, and difficulty in breathing, are associated with the phenanthridine alkaloids present in the lily family. If large quantities of the leaves and bulb are consumed, depression, ataxia, seizures, and hypotension may develop. Treatment if necessary should be directed towards relieving diarrhea and dehydration.

References
1. Martin SF: The Amaryllidaceae alkaloids. In The Alkaloids: Chemistry and Physiology. vol 30, Brossi A (ed) Academic Press, San Diego, Calif 251-376, 1987.
2. Watt JM, Breyer-Brandwijk MG: The medicinal and poisonous plants of southern and eastern Africa. 2nd ed. ItE & S Livingston, Edinburgh pp 33-39, 1962.

Family: Araliaceae

Figure 183 *Hedera helix* 'tricolor'

Figure 184 *Hedera helix*

Common Name
Ivy, English ivy *(Hedera helix)*, Algerian, Canary, or madeira ivy *(H. canariensis)*

Plant Description
A genus of about 10 species native to Europe, Asia, North Africa, and the Canary islands, *Hedera* species are evergreen, woody stemmed climbing vines. Ivies also make dense ground covers. Adventitious roots are produced along the stems that adhere the plant to the surface it is climbing. Leaves are simple, alternate, ovate or cordate, lobed or not, apices acute or obtuse, veination palmate, the upper surfaces glossy while the undersides are hairy or scaly (Figures 183 and 184). Inflorescences are generally umbels of flowers with 5 sepals and 5 petals. The fruits are black, yellow, or orange colored berries with 3-5 seeds.

Toxic Principle and Mechanism of Action
Hedera species contain triterpenoid saponins including hederasaponin B and C, hederasponoside B and C, caffeic acid, and hederin.[1] The leaves and fruits are toxic, and contain polyacetylene compounds such as falcarinol and dide-hydrofalcarinol that have been associated with irritant contact dermatitis in people who handle the plant excessively.[1-3]

Risk Assessment
Most problems with ivy are associated with contact dermatitis in some people who handle the plants and get the plant sap on their skin.[4,5] Similar contact dermatitis has not been reported in animals. However, ivy is a very popular plant in gardens and even as a potted house plant, and therefore is a potential risk to pets that might chew and eat the leaves and fruits.

Clinical Signs
The saponin content of the plants, are irritants, causing excessive salivation, vomiting, abdominal pain, and diarrhea. Symptoms are generally transient and treatment, if necessary, should be directed to relieving clinical signs.

References

1. Julien J. Gasquet M. Maillard C. Balansard G. Timon-David P. Extracts of the ivy plant, *Hedera helix*, and their anthelminthic activity on liver flukes. Planta Medica 53: 205-208, 1985.
2. Hausen BM, et al. Allergic and irritant contact dermatitis from falcarinol and didehydrofalcarinol in common ivy *(Hedera helix L.)*. Contact Dermatitis 17: 1-9, 1987.
3. Garcia M, Fernadez E, Navarro JA, del Pozo MD, Fernadez de Corres L. Allergic contact dermatitis from *Hedera helix* L. Contact Dermatitis 33: 133-134, 1995.
4. Lampe KF, McCann MA. AMA Handbook of Poisonous and Injurious Plants. Am Med Assoc. Chicago, Illinois 87-88, 1985.
5. Spoerke DG, Smolinske SC. Toxicity of House Plants. CRC Press. Boca Raton, Florida. 149-153, 1990.

Heliotropium

Family: Boraginaceae

Common Name

Heliotrope, cherry pie
The most common toxic species include blue heliotope *(H amplexicaule)*, common heliotrope *(H. arborescens)*, seaside or salt heliotrope *(H. currassavicum)*, and European heliotrope *(H. europaeum)*

Plant Description

There are over 250 species of *Heliotropium* occurring in most warm-temperate and tropical areas of the world. Annuals or perennials, herbs or shrubs, with rough, hairy or scaly, erect or prostrate stems. Leaves are generally opposite, lanceolate or ovate, sessile or petiolate, with basal leaves forming a rosette. Inflorescence is a terminal or axillary scorpioid cyme, or solitary flower with or without bracts. Flowers are small, funnelform, white, blue, or purple and rarely yellow in color (Figures 185 and 186). Fruits are nutlets, generally paired, irregular and coarsely wrinkled.

Figure 185 *Heliotropium* species

Figure 186 *Heliotropium* species

Toxic Principle and Mechanism of Action

Numerous toxic pyrrolizidine alkaloids are present in *Heliotropium* species, being more prevalent in the younger plants and the seeds.[1] Some of the more toxic alkaloids that have been isolated include lasiocarpine, heliotrine, europine, and heliosupine.[1,2] These alkaloids are converted by the liver into toxic pyrroles that inhibit cellular protein synthesis and cell mitosis.[1-3] Hepatocyte necrosis, fibrosis with biliary hyperplasia, and megalocytosis characterize the toxic effects of the pyrrolizidine alkaloids.[3] The chronic consumption of the alkaloids has a cumulative effect on the liver. Most animal poisoning has occurred in situations where the heliotrope is a dominant pasture weed, or when the plant or seeds contaminate hay or grain fed to livestock or poultry.[4-5] Sheep grazing heliotrope over long periods develop severe liver disease concurrently with copper poisoning.[7] Epidemics of *Heliotropium* poisoning characterized by severe hepatic veno-occlusive disease has occurred in people where bread has been contaminated with the seeds or where herbal medications contain the toxic alkaloids.[8,9]

Risk Assessment
Heliotrope poisoning is most frequently a problem in parts of the world where the plants grow in large populations that livestock have access to, or where the seeds contaminate grain fed to animals or birds. Some showy species of heliotrope *(H. arborescens)* are commonly grown as garden plants and its potential for poisoning should be recognized, especially where pruned clippings may be fed to livestock.

Clinical Signs
Animals that consume pyrrolizidine alkaloid containing plants over a period of time develop signs related to liver failure. Weight loss, icterus, diarrhea, photosensitization, and neurologic signs related to hepatic encephalopathy are typical of liver failure. Serum liver enzymes are generally elevated significantly.
Confirmation of PA toxicity can be made by a liver a biopsy showing the triad of histological changes characteristic of PA poisoning, namely liver megalocytosis, fibrosis, and biliary hyperplasia.[3]
Treatment of animals with PA poisoning is generally limited to placing the animal in a barn out of the sun to relieve the photosensitization, providing a high quality, low protein diet, and removing all sources of the PA from the animal's food. The prognosis is generally very poor as once clinical signs of liver failure from PA poisoning occur the degree of liver damage is severe and irreversible.

References
1. Mattocks AR: Chemistry and toxicology of pyrrolizidine alkaloids. Academic Press, Orlando Florida, 1986.
2. Cheeke PR; Toxicity and metabolism of pyrrolizidine alkaloids. J Anim Sci 66: 2343-2350, 1988.
3. Stegelmeier BL: Pyrrolizidine alkaloids. Clinical Veterinary Toxicology. Ed. Plumlee KH. Mosby St Louis, Missouri. 370-377, 2003.
4. Eroksuz Y, et al: Toxicity of dietary *Heliotropium dolosum* seed to broiler chickens. Vet Human Toxicol 43: 334-338, 2001.
5. Hill BD, Gaul KL, Noble JW: Poisoning of feedlot cattle by seeds of *Heliotropium europaeum*. Austral Vet J 75: 360-361, 1997.
6. Pass DA, et al: Poisoning of chickens and ducks by pyrrolizidine alkaloids of

Heliotropium europaeum. Austral Vet J 55: 284-288, 1979.
7. Howell JM, Deol HS, Dorling PR, Thomas JB: Experimental copper and heliotrope intoxication in sheep: morphological changes. J Comp Path 105: 49-74, 1991.
8. Tandon HD, Tandon BN, Mattocks AR: An epidemic of veno-occlusive disease of the liver in Afghanistan. Pathologic features. Am J Gastroenterol 70: 607-613, 1978.
9. Datta DV, Khuroo MS, Mattocks AR, Aikat BK, Chhuttani PN: Herbal medicines and veno-occlusive disease in India. Postgrad Med J 54: 511-515. 1978.

Helleborus

Family: Ranunculaceae

Common Names
Hellebore, Christmas rose *(H. niger)*, Lenten rose *(H. orientalis)*, stinking hellebore *(H. foetidus)*, stinkwort.

Plant Description
Hellebores are native to areas of Europe and Asia. At least 15 perennial or evergreen species are recognized for their showy flowers. Growing from rhizomatous or fibrous roots, hellebores have simple, alternate, palmate, or compound leaves with entire or toothed margins. The flowers are produced singularly or in small cymes with five or six large sepals that range from white green, purple, to red or yellow (Figures 187 and 188). After fertilization. the sepals turn green. The numerous smaller petals tend to fall off rapidly, and some are funneled-shaped and modified into nectaries. Numerous stamens are produced that are longer than petals. The fruits are beaked follicles.

Figure 187 *Helleborus niger*

Toxic Principle and Mechanism of Action
Helleborus species contain a variety of toxic compounds, including the irritant glycoside ranunculin that is converted to protoanemonin when the plant tissues are chewed and macerated.[1,2] Protoanemonin is a vesicant, and it is polymerized to the

Figure 188 *Helleborus argutifolius*

133

non-toxic anemonin. The dried plant contains mostly anemonin and is therefore not toxic. Additionally hellebores species also contain a number of cardenolides or bufadienolides that are cardiotoxic.[3] The roots are particularly toxic.

Risk Assessment
Hellebores are popular garden plants, especially because they are one of the earliest flowering plants in the spring. Although not reported as a cause of poisoning in household pets, the presence of significant quantities of cardenolides, and protoanemonins in these plants makes them potentially hazardous.

Clinical Signs
Excessive salivation, vomiting, and diarrhea can be anticipated if hellebores are eaten. In addition, the cardenolides can have profound effects upon the heart causing arrhythmias and heart block as might occur with digitalis poisoning. Treatment if necessary would be symptomatic.

References
1. Hill R, Van Heyningen R: Ranunculin: the precursor of the vesicant substance I Buttercup. Biochem J 49: 332-335, 1951.
2. Bonora A, Dall'Olio G, Bruni A: Separation and quantification of protoanemonins in Ranunculaceae by normal and reversed phase HPLC. Planta Med 51: 364-367, 1985.
3. Joubert JPJ: Cardiac glycosides. In toxicants of plant origin, vol 2, Glycosides.Cheeke PR (ed). CRC Press, Boca Raton, Florida pp 61-96, 1989.

Hemerocallis
Family: Liliaceae

Figure 189 *Hemerocallis* hybrid

Common Name
Day lilies

Plant Description
Hemerocallis is a genus of 15 species found primarily in Asia and Europe. Numerous hybrids have been made from the species. These perennial evergreen or semi-evergreen clump forming lilies are hardy under many growing conditions. The fragrant, star-shaped flowers are produced on stems above the foliage. The leaves are long, linear and in some species, grass-like. Flower size can range up to 6 inches in diameter, while flower color can range from white, yellow, orange, to red. Flowers only last a day, but day lilies are sequential bloomers. (Figures 189 and 190).

Toxic Principle and Mechanism of Action

Like the Easter lily, the tiger lily, the Asiatic or Japanese or Asiatic lilies, day lilies (*Hemerocallis* spp.) are toxic to cats causing nephrotoxicity that can prove to be fatal.[1-4] The toxin responsible for the nephrotoxicity of lilies has not been identified. All parts of the plants and especially the flowers, are poisonous to cats. Deaths have been reported in cats after ingestion of only two leaves.[3] Dogs, rats, and rabbits were not affected after they were experimentally fed high doses of Easter lily[3] (See Lilium species).

Figure 190 Day Lily

Risk Assessment

Day lilies and their hybrids are popular garden plants and are a significant risk to cats that are prone to chewing on plants.

Clinical Signs

Within 3 hours of consuming 1-2 leaves or flower petals of the lily, cats start to vomit and salivate excessively. The cats become depressed, anorexic, with the initial vomiting and salivation tending to subside after 4-6 hours. Approximately 24 hours later, proteinuria, urinary casts, isosthenuria, polyuria, and dehydration develop. Vomiting may recur at this stage. A disproportionate increase in serum creatinine as compared to blood urea nitrogen is a significant indicator of lily poisoning. As renal failure develops, anuria, progressive weakness, recumbency, and death occur.[1-4]

On postmortem examination the kidneys are swollen with perirenal edema. Mild to severe pulmonary and hepatic congestion is common. Renal tubular necrosis is the most prominent histopathologic lesion. Granular and hyaline casts are common in the collecting ducts.

Treatment

Cats suspected of eating lilies should be seen by a veterinarian as soon as possible after the plant was consumed. Gastrointestinal decontamination and fluid therapy is essential for preventing the nephrotoxicity. Vomiting, should be induced if the plant has been consumed within the past 1-2 hours. Activated charcoal with a cathartic should be given to decontaminate the gastrointestinal tract. Fluid therapy should be initiated to maintain renal function and prevent anuria. Fluid therapy should be continued for at least 24 hours. Once the cat has developed anuric renal failure, peritoneal dialysis or hemodialysis will be necessary to try and save the animal.

References

1. Hall JO: Nephrotoxicity of Easter lily *(Lilium longifolium)* when ingested by the cat. Proc Am Coll Vet Intern Med Forum 10: 840, 1992.
2. Carson TL, Sanderson TP, Halbur PG: Acute nephrotoxicity process in cats following ingestion of Lilium *(Lilium* spp.). Am Assoc Vet Lab Diagn 37: 43, 1994.
3. Hall JO: Lily. In clinical veterinary toxicology. Plumlee KH ed. Mosby, St. Louis, Missouri pp. 433-435, 2003.
4. Hadley RM, Richardson JA, Gwaltney-Brant SM: A retrospective study of daylily toxicosis in cats. Vet Human Toxicol 45: 38-39, 2003

Hibiscus

Family: Malvaceae

Figure 191 *Hibiscus schizopetalus*

Figure 192 *Hibiscus Rosa-Sinensis*

Common Names

Hibiscus, common rose mallow *(H. moscheutus)*, confederate or cotton rose *(H. mutabilis)*, rose of Sharon or blue hibiscus *(H. syriacus)*

Plant Description

Comprising a genus of about 220 species that are native in many tropical and subtropical areas of the world, *Hibiscus* species have been widely hybridized making many popular ornamental cultivars. Shrubs or small trees, with evergreen or deciduous, branching, woody stemmed, perennial or annuals. Leaves are simple, mostly smooth, toothed or not, and elliptic to cordate in shape. Flowers are produced terminally either singly or in spikes, and have 5 over-lapping, showy petals with a central column of fused stamens. Flower colors are numerous, varying from white to yellow and red (Figures 191-193).

Toxic Principle and Mechanism of Action

No toxin has been identified in *Hibiscus*, but dogs in particular that chew and eat the leaves develop excessive salivation, vomiting, diarrhea (often hemorrhagic) depression, anorexia, and dehydration.[1]

Risk Assessment

Hibiscus plants are common garden and household plants and therefore are potentially a problem to dogs that may chew on them.

Clinical Signs

Excessive salivation, vomiting, diarrhea that is often hemorrhagic, depression, anorexia, and dehydration may occur in dogs that eat the hibiscus plant parts.[1] Treatment is seldom necessary other than to relieve severe diarrhea and dehydration.

Figure 193 *Hibiscus* hybrid

References

1. Household Plant Reference. ASPCA National Animal Poison Control Center. 424 E. 92nd St., New York, NY 10128

Hippeastrum

Family: Liliaceae (Amaryllidaceae)

Common Names

Amaryllis, Barbados lily, naked lady, Azucena de Mejico, Resurrection Lily (Commonly referred to as amaryllis, *Hippeastrum* species are not a true amaryllis, as there is only one species of Amaryllis that is indigenous in southern Africa – see *Amaryllis belladonna*)

Plant Description

Originating in South America, the genus *Hippeastrum* has some 80 species of spectacular trumpet-shaped, tropical lilies that have become widely popular as seasonal potted plants, or as garden plants in tropical areas. Growing from large

Figure 194 Hippeastrum hybrid

fleshy bulbs, clusters of showy flowers are produced on hollow stems up 0.5 m in length often before the leaves emerge. Flower colors include white, pink, red, and red streaked with white (Figures 194 and 195). New hybrids are continually being developed. Normally blooming in the spring, *Hippeastrum* hybrids are frequently coaxed into blooming during the winter holiday season.

Toxic Principle and Mechanism of Action

Phenanthridine alkaloids including lycorine, haemanthamine, hippeastrine, tazettine, and vittatine have been identified in the leaves, stems, and bulbs of *Hippeastrum*.[1] The phenanthridine alkaloids are present in many of the Liliaceae,

Figure 195 *Hippeastrum* hybrid

Figure 196 *Hippeastrum* bulb

most notably in the *Narcissus* group (see *Narcisssus*). The alkaloids have emetic, hypotensive, and respiratory depressant effects, and cause excessive salivation, abdominal pain, and diarrhea. Calcium oxalate raphides in the plant tissues may also contribute to the digestive symptoms.

As many as 15 other phenanthridine alkaloids have been isolated from other genera of the Amaryllis family, including species of *Amaryllis, Clivia, Eucharis, Galanthus, Haemanthus, Hymenocallis, Leucojum, Narcissus, Nerine, Sprekelia, and Zephranthes.*[1]

Risk Assessment
This attractive garden and potted houseplant is one of the most popular plants, and therefore is commonly accessible to household pets. Usually purchased as a leafless bulb, and if not planted immediately, the bulb can be something puppies like to chew on (Figure 196).

Clinical Signs
Vomiting, excessive salivation, abdominal pain, diarrhea, and difficulty in breathing, are associated with the phenanthridine alkaloids present in the lily family. If large quantities of the leaves and bulb are consumed, depression, ataxia, seizures, and hypotension may develop. Poisoning is rarely fatal, and can generally be treated symptomatically.

References
1.Martin SF: The Amaryllidaceae alkaloids. In The Alkaloids: Chemistry and Physiology. vol 30. Brossi A (ed) Academic Press, San Diego, Calif pp 251-376, 1987.

Hosta

Family: Liliaceae

Common Names
Plantain lily, hosta lily, August lily

Plant Description
A genus of 40 species native to China and Japan, *Hosta* species have become universally popular for their decorative leaves. Numerous varieties have been developed. Frost hardy and shade tolerant perennials that produce large, showy leaves that are various shades of green, yellow, edged in yellow, or white. Inflorescences are tall racemes produced above the foliage. Flowers are bell-shaped, nodding, and may be white, bluish-purple, or pink in color (Figures 197 and 198).

Figure 197 *Hosta plantaginea*

Toxic Principle and Mechanism of Action
Little information is available on the toxicity of *Hosta* species. Some contain saponins that may produce vomiting, diarrhea, depression, and loss of appetite.[1,2]

Risk Assessment
Considered potentially toxic, hostas are only likely to be a problem if consumed in quantity.[2]

Clinical Signs
If hostas are consumed in quantity, the irritant effects of the saponins can be expected to cause vomiting and diarrhea.

Figure 198 *Hosta fortunei* 'Albomarginata'

References
1. Mimaki Y, Kuroda M, Kameyama A, Yokosuka A, Sashida Y: Steroidal saponins from the rhizomes of *Hosta sieboldii* and their cytostatic activity on HL-60 cells. Phytochemistry. 48: 1361-1369, 1998.
2. Household Plant Reference. ASPCA National Animal Poison Control Center. 424 E. 92nd St., New York, NY 10128.

Family: Asclepiadaceae

Figure 199 *Hoya carnosa compacta*

Figure 200 *Hoya carnosa*

Common Names
Hoya, wax plant, porcelain flower

Plant Description
Comprising a genus of over 200 species native to tropical Austral/Asia, *Hoya* are climbing, woody stemmed, evergreen plants with thick waxy, hairless, ovate to oblong, short petioled leaves. In some species the immature leaves are red before turning green. The characteristic star-shaped flowers are fragrant, produced in short stemmed umbels, and range in color from white to shades of pink and red (Figures 199 and 200).

Toxic Principle and Mechanism of Action
Cardenolides are present in all parts of the plants, and are mildly cardiotoxic. In Australia, *Hoya australis* has been reported to cause a neurologic syndrome in sheep and cattle characterized by incoordination, especially of the hind legs, tremors, and tetanic seizures.[1,2]

Risk Assessment
Hoyas are frost sensitive and therefore are primarily grown as house plants except in truly tropical areas. Poisoning of household pets from eating hoyas is rare. Hoyas are of minimal toxicity to canaries following experimental dosing.[3]

Clinical Signs
Cattle and sheep in Australia after eating quantities of *H. australis* develop neurologic signs including ataxia, tremors, tetanic-like seizures, and death depending on the quantity of plant consumed.

References
1. Everist SL: Poisonous plants of Australia. 2nd ed. Angus and Robertson, Sydney, Australia. 94-201, 1981.
2. Hungerford TG. Poisoning by plants. In Diseases of livestock 9th ed. Hungerford TG ed. 1624-1689, 1990.
3. Arai M, Stauber E, Shropshire CM: Evaluation of selected plants for their toxic effects in canaries. J Am Vet Med Assoc 200: 1329-1331, 1992.

Humulus lupulus

Family: Cannabidaceae

Common Names
Hops, beer hops, European hops, lupulin

Plant Description
Comprising a genus of 2 species, *Humulus* species are herbaceous perennials with twining hairy stems native to temperate zones of Europe and Asia. Native species found in North America are *H. neomexicanus*, *H. americanus*, and *H. japonicus*. Leaves are opposite, palmately 3-7 lobed, with ovate to round lobes. Cultivars have been developed that have variegated leaves. Male and female flowers are produced on separate plants in mid summer, the male flowers being small and in the leaf axils, while the female flowers are larger and cone-like. The female flowers turn yellow when mature and are used in brewing beer (Figures 201 and 202).

Figure 201 *Humulus lupulus*

Figure 202 *Humulus* female flower

Toxic Principle and Mechanism of Action
A variety of compounds are present in *Humulus* species including essential oils (humulene, myrcene, farnesene), phenolic compounds (coumaric, gallic, caffeic acid), resins, and biologically active proteins.[1,2] These compounds in themselves or their metabolites may be the source of fatal malignant hyperthermia reported in greyhounds.[3] It is possible that hops contain an uncoupler of oxidative phosphorylation that precipitates the hypothermia in greyhounds.[3] Hops also contain phytoestrogens that can effect the estrus cycle.[4]

Risk Assessment
It is unlikely that household pets will eat hops plants. Dogs however will readily eat spent hops from the brewing process. A cup full of the hops has been reported to cause poisoning in greyhounds. Spent hops should be carefully disposed of, and not thrown on the compost pile where dogs, especially grey hounds or greyhound crosses, could access them.

Clinical Signs
Marked hyperthermia, restlessness, panting, vomiting, signs of abdominal pain, and seizures occur within hours of greyhounds consuming spent hops. Creatinine

phosphokinase (CPK) levels become markedly elevated, and urine is dark brown suggesting muscle necrosis. Mortality can be high despite aggressive therapy for hyperthermia.

Dogs suspected of eating spent hops should be treated aggressively to empty the stomach and activated charcoal and a purgative administered. Intravenous therapy should be initiated to maintain renal function, and sedatives such as diazepam may help to control excitement and seizures.

References
1. Stevens R. The chemistry of hop constituents. Chem Rev 67: 19, 1967.
2. Westbrooks RG, Preacher JW. Poisonous plants of Eastern North America. University of South Carolina Press. 43, 1986.
3. Duncan KL, Hare WR, Buck WB. Malignant hyperthermia–like reaction secondary to ingestion of hops in five dogs. J Am Vet Med Assoc 210: 51-54, 1997.
4. Milligan SR, et al. Identification of a potent phytostrogen in hops *(Humulus lupulus)* and beer. J Clin Endocrin & Metabol 84: 2249-2252, 1999.

Hyacinthoides
Family: Liliaceae

Figure 203 *Hyacinthoides hispanica*

Common Names
English bluebell, harebell, wild hyacinth (Formerly considered *Scilla* or *Edymion*)

Plant Description
A genus of 4-6 species native to Europe and North Africa, *Hyacinthoides* species are bulbous perennials with basal linear to lanceolate, fleshy and concave leaves and a distinct flower stalk. The bulbs are globular and have onion-like papery outer layers. Inflorescences are terminal racemes, with 6-12 pendulous, bell-shaped, blue-purple, and rarely white flowers (Figure 203). Fruits are capsules with numerous black seeds.

Toxic Principle and Mechanism of Action
The bulbs and fruit capsules contain cardiotoxic cardenolides similar to bufadienolides found in squill (*Urginea* species). Other irritant compounds are likely present that cause gastrointestinal irritation. Poisoning has been reported in animals and humans who have eaten the bulbs mistaking them for onions.[1,2]

Risk Assessment
The bulbs pose the greatest risk to animals and children that might eat them.

Clinical Signs

Initially excess salivation, vomiting, and diarrhea may develop followed by depression, weakness, slow heart rate with dysrhythmias, and decreased cardiac output.[1,2]

Treatment is rarely necessary and should be directed at relieving gastrointestinal signs. Activated charcoal orally can be beneficial in preventing further absorption of the plant toxins.

References

1. Thursby-Pelham RHC. Suspected *Scill non-scripta* (bluebell) poisoning in cattle. Vet Rec 80: 709-710, 1967.
2. Burrows GE, Tyrl RJ: Toxic Plants of North America. Iowa State University Press, Ames. pp 769-770, 2001.

Hyacinthus

Family: Liliaceae

Common Names

Hyacinth, oriental hyacinth

Plant Description

Considered monotypic or a genus with 3 species, *Hyacinthus* are perennial, bulbous, spring flowering popular garden plants. Native to Asia, bulbs are globular, relatively large, with onion-like outer papery membranous layers. Leaves are glossy green, narrowly linear, and appear just before the flower stem. Inflorescence is an erect, terminal, cylindrical raceme with 2-40, waxy, strongly scented flowers. Each has 6 petals that are recurved or spreading, in shades of blue, pink or white (Figures 204 and 205). Fruits are 3 lobed, angular capsules with numerous winged seeds.

Figure 204 *Hyacinthus orientails*

Toxic Principle and Mechanism of Action

All parts of the plant and especially the bulbs contain similar toxins to tuliposide A and B found in tulips. A lectin and a glycoprotein have also been identified that may be responsible for the toxicity of tulips.[1,2]

Risk Assessment

Bulbs pose the greatest risk to people and animals that may handle or eat the bulbs. A contact dermatitis or allergy occurs in some individuals handling the tulip bulbs.[2-4]

Clinical Signs

Vomiting, increased salivation, increased heart rate, difficulty in breathing, and occasionally diarrhea can result. In cattle that have been fed tulip bulbs, intestinal

Figure 205 *Hyacinthus orientails* flowers

irritation, excessive salivation, decreased feed digestion, loss of weight, regurgitation of rumen contents, diarrhea, and death may result.[6,7]

Erythema, alopecia, and pustular lesions may develop in some people who handle the bulbs frequently. A similar contact dermatitis may also occur in some individuals who handle other common bulbs or plants of *Allstroemaria* species.[4]

References
1. Oda Y, Minami K: Isolation and characterization of a lectin from tulip bulbs, *Tulipa gesneriana*. Eur J Biochem 159: 239-246, 1986.
2. Spoerke DG, Smolinske SC: Toxicity of House Plants. CRC Press. Boca Raton, Florida. pp 212-214, 1990.
3. Hausen BM: Airborn contact dermatitis caused by tulip bulbs. J Am Acad Dermatol. 7: 500-504, 1982.
4. Hjorth N, Wilkinson DS: Contact dermatitis. IV. Tulip fingers, hyacinth itch, lily rash. Br J Dermatology 80: 696-69, 1968.
5. Maretic Z, Russel FE, Ladavac J: Tulip bulb poisoning. Period Biol 80: 141-143, 1978.
6. Wolf P. Blanke HJ. Wohlsein P. Kamphues J. Stober M: Animal nutrition for veterinarians--actual cases: tulip bulbs with leaves *(Tulipa gesneriana)*--an unusual and high risk plant for ruminant feeding. DTW - Deutsche Tierarztliche Wochenschrift. 110: 302-305, 2003.
7. Burrows GE, Tyrl RJ: Toxic Plants of North America. Iowa State University Press, Ames. 770-771, 2001.

Hydrangea

Family: Hydrangeaceae
(Saxifragaceae)

Common Names

Hydrangea, hortensia, hills of snow.
The most commonly cultivated species
are American hydrangea *(H. arborescens)*,
oak leaf hydrangea *(H. quercifolia)*, and
the large flowered Japanese hydrangea *(H. macrophylla)*.

Figure 206 *Hydrangea arborescens*

Plant Description

Deciduous or evergreen woody shrubs or
small trees, the 80 plus *Hydrangea* species
are native to a wide area of temperate
Asia and North and South America.
Leaves are large, ovate, with serrated
edges, and some with lobes such as the
oak leafed hydrangea. Inflorescences are
usually terminal panicles or heads with
numerous flowers. Flowers are radially
symmetrical, small and greenish-white.
The showy, nonfertile flowers have large
petal-like bracts ranging from white to
blue and pink in color. The color of the
bracts varies depending upon the pH of
the soil the plants are growing in. Acidic
soils produce blue-purple flowers, while
alkaline soils result in pink-red colors

Figure 207 *Hydrangea arborescens* 'lacecap'

(Figures 206-209). Hybrid inflorescences are made almost entirely of the infertile
flowers. Lacecap hydrangeas are those with a ring of non fertile flowers
surrounding the fertile ones (Figure 207).

Toxic Principle and Mechanism of Action

The cyanogenic glycoside, hydrangin, has been identified in the plant but its
significance in human and animal poisoning is unclear. The flower buds are
allegedly the most toxic.[1,2] A compound, isocoumarin hydrangenol, is also present
in the plant and is likely responsible for the contact dermatitis encountered in
some people who handle the plants.[2,3]

Risk Assessment

Other than the potential for causing dermatitis in some people, hydrangeas are
rarely associated with animal poisoning. None the less, the wilted or new leaves
have the potential for having significant cyanogenic glycoside that livestock
should not be fed the trimmings from the plants.

Figure 208 *Hydrangea macrophylla*

Figure 209 *Hydrangea macrophylla* 'Sir Joseph banks'

Clinical Signs

Vomiting, colic, diarrhea, and lethargy are signs that can be anticipated. Deaths are unlikely and animals appear to recover uneventfully. In people dermatitis of the hands and other parts of the body is reported.[1] Treatment is seldom necessary and is directed to- ward relief of symptoms.

References

1. Lampe KF, McCann MA: AMA Handbook of Poisonous and Injurious Plants. Am Med Assoc, Chicago, Illinois pp 94-95, 1985.
2. Spoerke DG, Smolinske SC: Toxicity of House Plants. CRC Press. Boca Raton, Florida pp 156-157, 1990.
3. Rademaker M. Occupational contact dermatitis to hydrangea. Australasian J Dermatol 44: 220-221, 2003.

Hymenocallis Species

(Synonym; Ismene)
Family: Liliaceae

Figure 210 *Hymenocallis narcissiflora*

Common Names

Spider lily, crown beauty, sea daffodil, sacred lily of the Incas, Peruvian daffodil, basket flower, filmy lily.

Plant Description

Consisting of about 40 species, related to the *Amaryllis* species, the *Hymenocallis* species are native to Central and South America, preferring areas that do not freeze. Ever green or deciduous, the strap-like leaves arise from bulbs, followed by the scented, showy, white flowers that resemble a daffodil except for the long slender petals surrounding the inner corona formed by the fused bases of the long anther filaments (Figures 210 and 211).
A similar genus, *Nerine*, consisting of 30 species of South African lilies with red, pink or white, narrow petals are popular garden plants in some tropical and

subtropical areas. These "spider" lilies (Guernsey lily) also do well as a potted plants indoors.

The genus *Leucojum* (Snowdrop) (Figure 212), with 10 species from North Africa and the Mediterranean area, is a bulbous lily with pendant, white, fragrant, bell-shaped flowers, and strap-like leaves. This species is similar in toxicity to *Hymenocallis* species.

Toxic Principle and Mechanism of Action

Several phenanthridine alkaloids including lycorine, and tazettine have been identified in the leaves, stems, and bulbs of *Hymenocallis* and *Nerine* species.[1-3] The phenanthridine alkaloids are present in many of the Liliaceae, most notably in the *Narcissus* group. The alkaloids have emetic, hypotensive, and respiratory depressant effects, and cause excessive salivation, abdominal pain, and diarrhea. Calcium oxalate raphides in the leaves and steams may also contribute to the digestive symptoms.

Risk Assessment

This attractive garden plant and occasionally potted houseplant has rarely caused poisoning in animals, but it has the potential to do so.

Figure 211 *Hymenocallis narcissiflora* (Peruvian daffodil)

Figure 212 *Leucojum vernum*

Clinical Signs

Vomiting, excessive salivation, abdominal pain, diarrhea, and difficulty in breathing, are associated with the phenanthridine alkaloids present in the lily family. If large quantities of the leaves and bulb are consumed, depression, ataxia, seizures, and hypotension may develop. Poisoning is rarely fatal and can generally be treated symptomatically.

References

1. Martin SF: The Amaryllidaceae alkaloids. In The Alkaloids: Chemistry and Physiology. vol 30, Brossi A (ed) Academic Press, San Diego, Calif 251-376, 1987.
2. Burrows GE and Tyrl RJ: eds. Liliacea. In Toxic Plants of North America. Iowa State University Press. Ames, Iowa pp 751-814, 2001.
3. Lin LZ et al: Lycorine alkaloids from *Hymenocallis littoralis*. Phytochemistry. 40: 1295-1298, 1995.

Hyoscyamus

Family: Solanaceae

Figure 213 *Hyoscyamus niger*

Figure 214 *Hyoscyamus niger* flower

Common Names
Henbane, black henbane, stinking nightshade

Plant Description
Comprising a genus of 15 species native to parts of Europe, North Africa, and Asia, *Hyoscyamus* species are erect, branching, annual, biennial, or perennial plants arising from a taproot and covered with prominent sticky hairs. Leaves are lanceolate to ovate, coarsely toothed, the lower leaves having petioles, the upper leaves clasping. The leaves form as a rosette the first year. Inflorescences are terminal one-sided, racemes or spikes. Flowers are showy, funnel shaped, 5-6 lobed, greenish to yellow in color with a distinct pattern of purple colored veins (Figures 213 and 214). Fruits are urn-shaped capsules enclosed by calyces, and containing many seeds (Figure 215).

Hyoscyamus niger (black henbane) is naturalized in North America and in some states is considered a noxious weed.

Toxic Principle and Mechanism of Action
All parts of the plant contain the tropane alkaloids hyoscyamine (atropine) and hyoscine (scopolamine). The tropane alkaloids are acetylcholine antagonists acting at the muscarinic cholinergic receptors in the autonomic nervous system, central nervous system, heart, and digestive systems, causing increased heart and respiratory rates, dilated pupils, excitement, intestinal stasis, and seizures. Neurotoxic calystegins are also present in *Hyoscyamus niger*.[1]

Risk Assessment
Henbane has been, and is a common cause of poisoning in people and has not been a problem to animals.[2-4] The plant has been used in herbal medicines for centuries for a variety of conditions and as a hypnotic, hallucinogenic, narcotic, and sedative.[5] Henbane is often grown as an ornamental for its interesting flowers, and the unique urn-shaped seed pods that are attractive in dry floral arrangements.

Consequently, the seeds are a potential hazard to children and pets.

Clinical Signs

Excitement, pupillary dilation, increased heart rate, labored breathing, colic, bloat, dry mouth, and seizures may be seen depending on the amount of plant ingested. Treatment is usually symptomatic and conservative. Activated charcoal orally as an adsorbent may reduce further absorption of the tropane alkaloids. Fluids and electrolytes intravenously may be necessary in severely intoxicated animals. Seizures may be controlled by diazepam, and in severe cases physostigmine may be used as a short acting cholinergic to reverse the atropine effects on the autonomic nervous system.

Figure 215 *Hyoscyamus niger* seed capsules

References

1. Drager B, van Almsick A, Mrachatz G. Distribution of calystegins in several Solanaceae. Planta Med 61: 577-579, 1995.
2. Daneshvar S, Mirhossaini ME, Balali M. *Hyoscyamus* poisoning in Mashhad. Toxicon 30: 501, 1992.
3. Spoerke DG, et al.BH. Mystery root ingestion. J Emerg Med. 5: 385-388, 1987.
4. Manriquez O, Varas J, Rios JC, Concha F, Paris E. Analysis of 156 cases of plant intoxication received in the Toxicologic Information Center at Catholic University of Chile. Vet Hum Toxicol 44: 31-32, 2002.
5. Duke JA. Handbook of Medicinal Herbs. 6th Ed. CRC Press Inc. Boca Raton, Florida 240-241, 1988.

Ilex

Family: Aquifoliaceae

Common Names

English, Chinese, Japanese holly, inkberry, gallberry

Plant Description

The 400 plus species of the genus *Ilex* are found throughout the temperate regions of the world; 29 species are native to North America. Generally shrubs or trees, deciduous or evergreen, attaining heights to 15m. Leaves are glossy, leathery, often with spiny edges. Flowers are generally

Figure 216 *Ilex aquifoluim* 'aurea marginata'

Figure 217 *Ilex cornuta*

Figure 218 *Ilex aquifoluim*

small and white, greenish-white or yellow in color. The characteristic fruits range in color from bright red to yellow or black depending on the species (Figures 216-218).

Toxic Principle and Mechanism of Action

A variety of compounds are found in *Ilex* species, including saponins, methylxanthines (caffeine, theobromine), and cyanogenic glycosides.[1] The saponins are however, the primary toxicants responsible for the gastrointestinal disturbances associated with eating the leaves or berries.

Risk Assessment

Branches of holly containing bright red berries are frequently used for decorative purposes during the winter holiday season in many households. Consumption of the berries by children and household pets is a frequent concern.[2] However, unless the berries are eaten in considerable quantity there is little risk, other than the possibility of some vomiting and diarrhea. In one study, holly was fed to cattle with no deleterious effect.[3]

Clinical Signs

Clinical signs of Ilex poisoning are generally limited to mild to moderate vomiting and diarrhea. Activated charcoal given orally is often sufficient to prevent gastrointestinal irritation. When diarrhea is severe, fluids and electrolytes should be administered as necessary.

References

1. Krenzelok EP, Jacobsen TD: Plant exposures... a national profile of the most common plant genera. Vet Human Toxicol 39: 248-249, 1997.
2. Alikaridis F. Natural constituents of *Ilex* species. J Ethnopharmacol 20: 121-144, 1987.
3. Pence M. Frazier KS. Hawkins L. Styer EL. Thompson LJ. The potential toxicity of *Ilex myrtifolia* in beef cattle. Vet Human Toxicol. 43: 172-4, 2001.

Ipomoea

Family: Convolvulaceaea

Common Name
Morning glory

Plant Description
This large genus of some 500 species of climbing, annual or perennial, vines or shrubs, are cosmopolitan to warm-temperate and tropical areas. Leaves are simple or palmately lobed, often with long twinning petioles. Flowers are large, showy, funnel-shaped, varying in color from white, purple, blue and pink to red depending upon the species and cultivar (Figures 219 and 220). Fruits are ovoid capsules containing 1-4 seeds.

Figure 219 *Ipomoea leptophylla*

Toxic Principle and Mechanism of Action
The seeds in particular contain high levels of indole alkaloids similar to the ergot alkaloids. The primary alkaloids in the seeds of *Ipomoea tricolor* are lysergic acid, isolysergic acid, and chanoclavine.[1] Varieties of *I. tricolor* containing lysergic acid include "Heavenly Blue," "Pearly gates," "Summer skies," "Blue star," and "Wedding bells."[2] Not all species of

Figure 220 *Ipomoea tricolor*

Ipomoea contain the lysergic type alkaloids, but instead have various other non-toxic alkaloids. Some of the species grown as garden ornamentals with low lysergic acid content include *I. alba* (Moon vine), *I. coccinea* (red morning glory), *I. nil* (Scarlet O'Hara), and *I. xsloteri* (cardinal climber).[3] Depending on the quantity of seeds that are well chewed and swallowed, the effects vary from marked hallucinations, incoordination, lethargy, diarrhea, to, in rats, increased mortality.[4,5]

Species of *Ipomoea* from Africa, and Australia *(I. carnea)* have been shown to contain the alkaloids swainsonine and calesteings that inhibit the cellular enzyme mannosidase and cause lysosomal storage disease.[6,7] These same alkaloids are found in other species including *Astragalus, Oxytropis* (locoweeds), *Convolulus* (bindweed) and *Calestegia* (hedge bindweed).[8] These exotic species of morning glory are also capable of causing anemia and hepatic necrosis in ruminants that eat the leaves and seeds.[9]

Moldy sweet potatoes *(I. batatas)* contain furan compounds such as ipomeanol that are known to cause severe respiratory disease in cattle characterized by destruction of type I pneumocytes in the lungs. Type II pneumocytes then proliferate causing interstitial pneumonia.[10]

Risk Assessment
Poisoning of dogs and cats is unlikely as many seeds of the morning glory have to be well chewed and swallowed. A mild toxic dose in humans is considered 20-50 seeds.[2]

Clinical Signs
In the rare circumstance where a toxic dose of the lysergic acid containing morning glory seeds was ingested, the affected animal could be expected to show signs varying from lethargy, incoordination, abnormal behavior, and diarrhea. Treatment in such cases might require the use of emetics and purgatives to remove the seeds from the digestive tract. Emesis should be used with caution in a very depressed or comatose animal.

References
1. Taber WA, Vining LC, Heacock RA. Clavine and lysergic acid alkaloids in varieties of morning glory. Phytochemistry 2: 65-70, 1963.
2. Der Mardersonian A. Psychotomimetic indoles in the Convolvulaceae. Am J Pharm 139: 19-26, 1967.
3. Wilkinson RE, Hardcastle WS, McCormick CS. Seed alkaloid contents of *Ipomoea hederifolia, I. quamoclit, I. coccinea, and I. wrightii.* J Agric Food Sci 198, 39: 335-339.
4. Dugan GM, Gumbmann MR. Toxicologic evaluation of morning glory seed: sub-chronic 90 day feeding study. 17: 647-675, 1977.
5. Rice WB, Genest K. Acute toxicity of extracts of morning glory seeds in mice. Nature 207: 302-303, 1965.
6. Ikeda K. Kato A. Adachi I. Haraguchi M. Asano N. Alkaloids from the poisonous plant *Ipomoea carnea*: effects on intracellular lysosomal glycosidase activities in human lymphoblast cultures. J Agri & Food Chem. 51: 7642-7646, 2003.
7. Haraguchi M, et al. Alkaloidal components in the poisonous plant, *Ipomoea carnea* (Convolvulaceae). J Agric & Food Chem. 51: 4995-5000, 2003.
8. Molyneux RJ, James LF, Ralphs MH, Pfister JA, Panter KE, Nash RJ. Polyhydroxy alkaloid glycosidase inhibitors from poisonous plants of global distribution: analysis and identification. In Plant-Associated Toxins: Agricultural, Phytochemical, and Ecological Aspects. Colegate SM, Dorling PR eds. CAB International, Wallingford, UK 107-112, 1994.
9. Damir SA, Adam SEI, Tartour G. The effects of *Ipomoea carnea* on goats and sheep. Vet Hum Toxicol 29: 316-319, 1987.
10. Wilson BJ, Burka LT. Sweet potato toxins and related toxic furans. In Handbook of Natural Toxins Vol 1. Plant and Fungal Toxins, Keeler RF, Tu AT eds. Marcel Dekker, New York 3-41, 1983.

Iris

Family: Iridaceae

Common Names
Iris, sword lily, blue or yellow flag

Plant Description
A genus of about 300 species of upright perennials arising from a tuberous root or bulb that are native to the northern hemisphere. Leaves are basal, leathery, linear, often arranged in a fan, and are deciduous in many species. Flowers are produced in terminal clusters of one to several, and are radially symmetrical. Some species have a ridge of hairs along the center of the petals and are referred to as bearded irises. Iris species have 3 sepals and 3 petals, and come in colors of white, blue, yellow, and purple (Figures 221-224). Numerous hybrids have been developed. Fruits are oblong, 3-6 angled capsules with numerous seeds arranged in rows.

Figure 221 *Iris reticulata*

Figure 222 *Iris pseudacorus*

Toxic Principle and Mechanism of Action
All parts of the plant but especially the rhizomatous roots, bulbs, and seed capsules contain pentacyclic terpenoids such as missourin, missouriensin, and zeorin.[1] These compounds are irritants and cause gastroenteritis if ingested, and contact dermatitis in some people handling and contacting the sap.[2,3]

Risk Assessment
Irises are commonly grown for their striking flowers and the rhizomatous roots, which grow at the soil surface, are readily accessible to pets. Rhizomes or bulbs stored over-winter in the house are always a potential source of poisoning to household pets.

Figure 223 *Bearded Iris* hybrid

Figure 224 *Beardless Iris*

Clinical Signs

Excessive salivation, vomiting, and diarrhea follow the ingestion of iris rhizomes. Treatment when necessary may require intestinal protectants and intravenous fluids to counteract the effects of dehydration.

References

1. Connolly JD, Hill RA: Dictionary of Terpenoids Vol 2. Di- and higher terpenoids. Chapman & Hall, London. 1382-1501, 1991.
2. Spoerke DG, Smolinske SC: Toxicity of House Plants. CRC Press. Boca Raton, Florida. 159-161, 1990.
3. Lampe KF, McCann MA: AMA Handbook of Poisonous and Injurious Plants. Am Med Assoc, Chicago, Illinois 98, 1985.

Jatropha

Family: Euphorbiaceae

Figure 225 *Jatropha podagrica*

Common Names

Barbados nut, physic nut, purging nut, coral plant

Plant Description

This large genus of some 200 species is native to the warm temperate and tropical regions of Asia, the Americas, and especially South America. Some of the species are desirable ornamentals because of their large attractive leaves and bright and unique flowers. Some species tolerate dry conditions and do well in cactus gardens.

Jatropha species can be perennial herbs, shrubs, or small trees. Most have a milky sap. Stems are erect, woody or fleshy especially basally. Depending on the species, some are monoecious, and others dioecious. The leaves are simple, alternate, or fascicled, and with varying shapes including being pinnate, palmate, chordate, or reniform. The margins are entire or toothed. Inflorescences are compound small clusters or cymes produced terminally or from leaf axils. Individual flowers are small, with 5 sepals and 5 petals often fused with the adjacent gland. Colors range from bright red to purple and yellow. Fruits are capsules containing 1-3 seeds (Figures 225-228).

Toxic Principle and Mechanism of Action

All parts of the plant, and in particular, the seeds contain a wide variety of diterpenoid esters of the tigliane type similar to those found in other members of the euphorbia family.[1] The primary effect of the diterpenoid esters is gastrointestinal irritation characterized by excessive salivation, vomiting, abdominal pain, and diarrhea. Various other compounds are present in the seeds, but have no or unknown effects on animals. A lectin, similar to the ricin in *Ricinus communis*, is also present in the seeds and may contribute to the gastrointestinal irritation.[2]

Figure 226 *Jatropha podagrica* flowers and fruits

Risk Assessment

Under most circumstances *Jatropha* species are not a particular problem to household pets or children as they are relatively unusual house or garden plants. However their popularity in tropical areas is increasing and therefore the potential risk of the seeds being ingested is a factor. Most poisoning occurs in children who eat the seeds.[3,4] The toxicity of some species of *Jatropha* has been well documented in sheep, goats, cattle, and

Figure 227 *Jatropha multifida*

chickens.[5-7] The lethal dose of *Jatropha curcas* seeds in calves was as low as 0.25gm/kg body weight.[7]

Clinical Signs

The irritant effects of the *Jatropha* are usually manifested within a few hours of ingestion of the seeds and consist of excessive salivation, vomiting, and severe diarrhea that can lead to dehydration and death. Treatment consists of aggressive fluid therapy, and intestinal protectants to counter the dehydration from the diarrhea. The major pathologic findings include generalized enteritis, hemorrhages through many organs, hepatic degeneration, and pulmonary congestion.[7]

References

1. Connolly JD, Hill RA:Dictionary of Diterpenoids vol2.Dip and higher terpenoids. Chapman & Hall, London pp 1277-1279, 1991.
2. el Badwi SM, Mousa HM, Adam SE, Hapke HJ: Response of brown hisex chicks to low levels of *Jatropha curcas, Ricinus communis* or their mixture. Vet Human Toxicol. 34: 304-6, 1992.

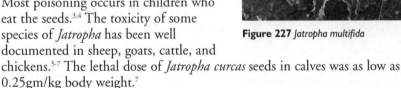

Figure 228 *Jatropha integerrima*

3. Levin Y, Sherer Y, Bibi H, Schlesinger M, Hay E: Rare *Jatropha multifida* intoxication in two children. J Emergency Med. 19: 173-5, 2000.
4. Abdu-Aguye I, Sannusi A, Alafiya-Tayo RA, Bhusnurmath SR: Acute toxicity studies with *Jatropha curcas* L. Human Toxicol 5: 269-274, 1986
5. el Badwi SM, Adam SE: Toxic effects of low levels of dietary *Jatropha curcas* seed on Brown Hisex chicks. Vet Human Toxicol. 34: 112-5, 1992
6. Ahmed OM, Adam SE: Toxicity of *Jatropha curcas* in sheep and goats. Res Vet Sci 27: 89-96, 1979.
7. Ahmed OM, Adam SE: Effects of *Jatropha curcas* on calves. Vet Path. 16: 476-82, 1979.

Juglans

Family: Juglandaceae

Figure 229 *Juglans nigra*

Common Name
Black walnut

Plant Description
A genus of some 20 species of hardwood trees native to Europe, Asia, and North America, *Juglans* species grow from 3-50m in height, and are valued for their wood and edible nuts. These large branching trees have a gray-brown ridged bark and leaves that are alternate, pinnately compound, deciduous and with 5-25 leaflets. Male and female flowers are produced on separate trees; the male or staminate flowers are catkins up to 12 cm in length, while the female or pistillate flowers 1-3cm long with yellow-green stigmas. The heart wood is characteristically brown to purple in color. The fruits are ovoid, hard-shelled nuts containing edible fruit (Figures 229-231).

Other common species of walnut with edible fruits include:
Juglans regia – English walnut, Persian walnut
J. cineraria – butternut, white walnut

Toxic Principle and Mechanism of Action
Horses exposed to black walnut shavings may develop varying degrees of laminitis.[1,2] The toxic component of black walnut responsible for causing laminitis

in horses has not been determined, but it is not apparently juglone, a naphthaquinone, which has been shown experimentally to cause lung and liver necrosis in dogs, and is fatal to fish.[3,4] From this research, it would appear that juglone has toxic effects on cell membranes.[3] Similar experimental intravenous administration of juglone to ponies produced pulmonary edema, possibly as a result of an anaphylactic response to the juglone.[2] The heartwood of black walnut does not contain juglone, while the leaves contain as much as 150 ppm.[5,6] Aqueous extracts from the heartwood of black walnut when administered intravenously to horses does cause laminitis apparently as a result of vasoconstriction of the vessels supplying the sensitive laminae of the hoof.[7,8]

Figure 230 *Juglans migrans* fruits

Horses bedded on fresh wood shavings containing as little as 5-20% walnut shavings can develop laminitis.[9-11] It appears that horses must obtain the toxin orally, or possibly from inhalation because fresh walnut shavings applied to directly to the horses hooves experimentally failed to cause laminitis.[2] Weathering of walnut shavings appears to reduce their toxicity.

Figure 231 Black walnuts

Black walnuts that have over wintered on the ground and that have become moldy are a source of tremorgenic mycotoxins (Penitrem A) produced by *Penicillium* species that can cause muscle tremors and even seizures in dogs that eat the moldy walnuts.[12,13]

Risk Assessment
New wood shavings containing walnut shavings should not be used for horse bedding.

Household pets are not likely to be exposed to walnut shavings. Well weathered walnut shavings if used in dog pens are not likely to be toxic. Dogs should not be allowed to eat moldy walnuts because of the potential for mycotoxins being present.[13]

Clinical Signs
Horses that have been bedded on fresh walnut shavings may develop elevated body temperature, increased heart and respiratory rates, edema of the legs, marked digital pulse, and lameness due to laminitis.[2] Ponies and foals appear to be less affected than adult horses, but signs are quite variable.[2] Signs of laminitis may appear from 1-3 days after exposure to the walnut shavings.

Treatment consists of immediately removing the horse from the walnut shavings and placing it in a sand stall or one with rubber flooring. Activated charcoal given orally via nasogastric tube may be beneficial if a horse is thought to have eaten shavings. It may also be prudent to wash the horse to remove all possible sources of walnut from the animal. Supportive treatment including phenylbutazone, flunixin meglumine, and appropriate hoof care administered as necessary can result in complete recovery from the laminitis.[2] In severe cases, the laminitis can progress to the point that there is separation and rotation of the third phalanx and may result in severe lameness requiring extensive veterinary care.

Dogs that have eaten moldy walnuts may develop muscle tremors, weakness, and seizures generally recover once the source of the mycotoxin is removed.[13]

References

1. Uhlinger C: Black walnut toxicosis in ten horses. J Am Vet Med Assoc 195:343-344, 1989.
2. True RG, Lowe JE: Induced juglone toxicosis in ponies and horses. Am J Vet Res 41: 941-945, 1980.
3. Boelkins JN, Everson LK, Auyoung TK: Effects of intravenous juglone in the dog. Toxicon 6: 99-102, 1968.
4. Marking LL: Juglone (5-hydroxy-1,4-napthaquinone) as a fish toxicant. Transcripts Am Fish 99: 510-512, 1979.
5. Coder KD: Seasonal changes of juglone potential in leaves of blackwalnut (*Juglans nigra* L) J Chem Ecol 9: 1203-1206, 1983.
6. Minnick PD et al: The induction of equine laminitis with an aqueous extract of the heartwood of black walnut (*Juglans nigra*). Vet Hum Toxicol 29: 230-233, 1987.
7. Galey FD, Whiteley HE, Goetz TE, Kuenstler AR, Davis CA, Beasley VR: Black walnut (*Juglans nigra*) toxicosis: a model for equine laminitis. J Comp Pathol 104:313-326, 1991.
8. Galey FD, Beasley VR, Schaeffer D, Davis LE: Effect of an aqueous extract of black walnut (*Juglans nigra*) on isolated equine digital vessels. Am J Vet Res 51: 83-88, 1990.
9. Galey FD: In Clinical Veterinary Toxicology. Plumlee KH ed. Mosby Inc. St Louis, Missouri, 425-427, 2004.
10. Ralston SL, Rich VA: Black walnut toxicosis in horses. J Am Vet Med Assoc 183: 1095, 1983.
11. Thomsen ME. Davis EG. Rush BR. Black walnut induced laminitis. Vet Human Toxicol 42: 8-11, 2000.
12. Abdel-Hafez AI. Saber SM. Mycoflora and mycotoxin of hazelnut (*Corylus avellana* L.) and walnut (*Juglans regia* L.) seeds in Egypt. Zentralblatt fur Mikrobiologie. 148(2): 137-47, 1993.
13. Richard JL et al. Moldy walnut intoxication in a dog caused by the mycotoxin penitrem A. Mycopathologica 76: 55-58, 1981.

Kalanchoe

(Synonym: Bryophyllum)
Family: Crassulaceae

Common Names
Kalanchoe, Palm Beach bells, flaming Katie, panda plant, pussy ears

Plant Description
There are approximately 200 species of *Kalanchoe* native to Africa, Madagascar, Asia, Arabia, and tropical America. This diverse genus consists of herbs, subshrubs, and climbers and are tropical perennial plants characterized by fleshy steams with fleshy leaves or inflorescences. The erect, often woody, plants attain heights of 3 m depending on the species. Leaves may be opposite or whorled, sessile or petiolate, linear to broadly ovate, succulent, hairless, or hairy. Inflorescences are panicles with few to many tubular or bell shaped flowers in colors of white, yellow, orange, brown, red, or purple (Figures 232-236). The fruits are many seeded capsules or follicles.

The most common kalanchoe sold by florists for household use is *Kalanchoe blossfeldiana* and its hybrids (Figures 235 and 236).

Toxic Principle and Mechanism of Action
A variety of cardiotoxic bufadienolides are present in all parts of the plant, but especially in the flowers.[1,2] The cardiotoxins, including bryotoxins, bryophyllins, and bersalgenins, have similar action to the cardiac glycosides found in foxglove, oleander, Lily of the valley, and milkweeds. Their primary effect is to inhibit Na^+/K^+ adenosine triphosphatase, thereby decreasing the transportation of sodium and potassium

Figure 232 *Kalanchoe thyrsiflora*

Figure 233 *Kalanchoe prolifera*

Figure 234 *Kalanchoe daigremontiana*

across cell membranes which decreases cardiac function. The most toxic species is *K. delagonensis* and its hybrids.[3] Although there is considerable variation in the toxicity of the Kalanchoe species, all species should be considered toxic until proven otherwise.

Most cases of poisoning reported in animals have occurred in cattle, but other animals including birds are susceptible.[2-6]

A similar African genus, *Cotyledon* contains cardiotoxic bufadienolides, and has been associated with poisoning in sheep.[7] The most commonly encountered toxic species in North America is *C. orbiculata* (pigs ears) (see Crassula).

Risk Assessment

Kalanchoes are an increasingly popular florist plant in North America, and therefore as potted houseplants pose a significant risk to household pets and children that may eat the plants. In tropical areas, kalanchoes are popular garden plants as they are drought tolerant.

Figure 235 *Kalanchoe blossfeldiana* hybrid

Two genera of plants that resemble the kalanchoes are the jade plants (*Crassula* species) and the *Sedum* species. These species are common garden and house plants, and although quite frequently the source of concern, there is no evidence that they are very toxic but may occasionally cause some salivation and vomiting if ingested.

Figure 236 *Kalanchoe blossfeldiana* hybrid

Clinical Signs

Animals consuming kalanchoe become depressed, anorexic, and develop excessive salivation and persistent diarrhea within a few hours of consuming the plants. Cardiac dysrhythmias, bradycardia, and heart block may develop and can result in death acutely or after several days. Blood urea nitrogen, creatinine, and blood glucose levels are frequently increased.

The oral administration of activated charcoal shortly after the plant is consumed may help decrease consumption of the toxins. Supportive treatment with fluids and electrolytes to counteract the effects of the persistent diarrhea are indicated. Where cardiac dysrhythmias are severe enough to affect cardiac function, atropine or propranolol may be indicated.

References

1. Oelrichs PB, MacLeod JK, Capon RJ: Isolation and identification of the toxic principles in *Bryophyllum tubiflorum (Kalanchoe)*. In Poisonous Plants: Proceedings of the third international symposium.James LF, Keeler RF, Bailey EM, Cheeke PR eds. Iowa State press, Ames pp 288-292, 1992.
2. McKenzie RA, Dunster PJ: Hearts and flowers: Bryophyllum poisoning of cattle. Austr Vet J 63: 222-227, 1986.
3. McKenzie RA, Franke FP, Dunster PJ: The toxicity to cattle and bufadienolide content of six *Bryophyllum* species. Austr Vet J 64: 298-301, 1987.
4. Masvingwe C, Mavenyengwa M: *Kalanchoe lanceolata* poisoning in Brahman cattle in Zimbabwe: the first field outbreak. J South African Vet Assoc 68: 18-20, 1997.
5. Williams MC, Smith MC: Toxicity of *Kalanchoe* spp to chicks. Am J Vet Res 45: 543-546, 1984.
6. Anderson LA et al: Krimpsiekte and acute cardiac glycoside poisoning in sheep caused by bufadienolides from the plant *Kalanchoe lanceolata* Forsk. Onderstepoort J Vet Res 50: 295-300, 1983.
7. Kellerman TS, Coetzer JAW, Naude TW: Plant Poisonings and Mycotoxicoses of Livestock in Southern Africa. Oxford University Press, Capetown pp 99-108, 1990.

Kalmia

Family: Ericaceae (Heath family)

Common Names
Laurel, mountain laurel, sheep laurel, sheepkill, calico bush

Plant Description
Native to the acidic swampy soils of the temperate regions of North America, the 6-10 species of *Kalmia* are small to large, erect, branching shrubs, with grey to reddish bark that exfoliates as the plant matures. Leaves may be deciduous or evergreen, alternate, opposite or whorled,

Figure 237 *Kalmia latifolia*

glabrous, dark green, revolute, and elliptic to lanceolate in shape. Inflorescences are terminal or axillary corymbs. The flowers are showy white, pink, or red with 5 fused sepals and 5 fused petals. Each petal is keeled forming a total of 10 pouches, an anther positioned in each pouch. Fruits are ovoid capsules (Figures 237 and 238).

The species of laurel most frequently associated with toxicity include *Kalmia angustifolia* (sheep laurel), *K. latifoli* (mountain laurel), *K. microphylla* (alpine laurel), *K. polifolia* (swamp laurel). Numerous cultivars have been developed, making laurels popular garden plants.

Figure 238 *Kalmia latifolia* cultivar

Toxic Principle and Mechanism of Action

All species of the family Ericaceae contain varying quantities of toxic diterpenoids collectively known as grayanotoxins I and II (formerly andromedotoxin, rhodotoxin, and acetylandromedol).[1] As many as 18 grayanotoxins (I – XVIII) have been identified, the greatest number being found in the *Leucothoe* species (fetter bush).[2,3] Tannins and other compounds are also present in varying amounts. All parts of the laurel including the flowers are toxic, although there may be considerable variation between species.

Grayanotoxins act to increase sodium channel permeability of cells by opening the channels to sodium which enters the cells in exchange for calcium ions, thus rendering the channels slow to close so that the cell remains depolarized.[4,5] Other neurologic mechanisms may also involve a cholinergic response seen clinically as bradycardia and excessive salivation.[6] The cardiac effects can range from bradycardia, sinus arrest, to arrhythmias.

Other members of the Ericaceae that contain grayanotoxins include:

Andromeda polifolia	Andromeda, bog rosemary
Ledum spp.	Labrador tea
Leucothoe spp.	Fetter bush, dog laurel
Lyonia spp.	Maleberry
Menziesia spp.	Rusty menziesia
Pieris spp.	Pieris, Japanese pieris
Rhododendron spp.	Rhododendrons, azaleas

Risk Assessment

Laurel species are commonly grown as showy garden shrubs. Livestock poisoning occurs where laurels are accessible to the animals. Laurels are potentially toxic to all animals.

Clinical Signs

Excessive salivation, increased nasal secretions, vomiting, abdominal pain, bloat, and irregular respirations develop several hours after rhododendron leaves are ingested.[8,9] Projectile vomiting may be noticeable. Hypotension, tachycardia, dysrhythmias, and respiratory depression may also develop. Weakness, partial blindness, and seizures have been reported in severe intoxications. Neurologic signs may persist for several days before the animal recovers. Weight loss may be notable.

Treatment is primarily directed at relief of the more severe clinical signs. Activated charcoal given orally is helpful if given shortly after the laurel is consumed. Atropine is useful in countering the cardiovascular effects

References
1. Kakisawa H, Kozima T, Yanai M, Nakanishi K: Stereochemistry of grayanotoxins. Tetrahedron 21: 3091-3104, 1965.
2. Sakikabara J, Shirai N, Kaiya T, Nakata H: Grayanotoxin-XVIII and its grayanoside B, a new A-Nor-B-Homo-Ent-Kaurine and its glucoside from *Leucothoe grayana.* Phytochemistry 18: 135-137, 1979.
3. Sakikabara J, Shirai N, Kaiya T: Diterpene glycosides from *Pieris japonica.* Phytochemistry 20: 1744-1745, 1981.
4. Narahashi T, Seyama I: Mechanism of nerve membrane depolarization caused by grayanotoxin I. J Physiol 242: 471-487, 1974.
5. Seyama I, Narahashi T: Modulation of sodium channels of squid nerve membranes by grayanotoxin I. J Pharmacol Exp Ther 219: 614-624, 1981.
6. Onat F, Yegan BC, Lawrence R, Oktay A, Oktay S: Site of action of grayanotoxins in mad honey in rats. J Appl Toxicol 11: 199-201,1991.
7. Krochmal C: Poison honeys. Am Bee J 134: 549-550, 1994.
8. Casteel SW, Wagstaff JD: *Rhododendron macrophyllum* poisoning in a group of goats and sheep. Vet Hum Toxicol 31: 176-178, 1989.
9. Puschner B et al: Azalea (*Rhododendron* spp.) toxicosis in a group of Nubian goats. Proc 42nd Annual Meeting Am Assoc Vet Lab Diagn 1999.

Laburnum anagyroides

Family: Fabaceae

Common Names
Laburnum, golden chain tree, golden rain tree

Plant Description
Originating in southern Europe, the genus *Laburnum* has 2 species, *L. anagyroides and L. alpinum,* that have been used to produce the popular hybrid *L. watereri* seen commonly in cultivation. *Laburnum* species are perennial, deciduous, branching shrubs or small trees with smooth grayish-green bark. Leaves are palmate, compound with 3 leaflets. Inflorescences are long (15-30cm) pendulous racemes produced from

Figure 239 *Laburnum anagyroides*

leaf axils. The showy yellow pea-like flowers have 5 fused sepals, 5 petals, the banner being rounded to obovate, the keel convex and shorter than the wing petals (Figures 239 and 240). Fruits are linear legume pods with flat brown seeds.

Toxic Principle and Mechanism of Action
The primary toxicants in *Laburnum* species are the quinolizidine alkaloids cytisine and N-methylcytisine. The teratogenic quinolizidine alkaloid anagyrine is also present along with a variety of others in smaller amounts. Although present

Figure 240 *Laburnum watereri*

in all parts of the plant, the greatest concentrations of the alkaloids occur in the seeds. Most cases of poisoning are associated with consumption of the pods and seeds. Horses appear to be most sensitive to the alkaloids, but poisoning has been reported in cattle, dogs, pigs, and humans.[1-3] The toxic dose of seeds in horses has been estimated to be 0.5mg/kg body weight. Cattle appear to tolerate considerably more seed with signs appearing when 30.5mg/kg body weight is fed.[2]

Cystisine is rapidly absorbed and excreted and consequently clinical signs of poisoning occur rapidly after a toxic dose of the seeds are consumed. Equally, the signs are relatively short – lived due to rapid excretion of the alkaloid. Cytisine binds strongly to nicotinic receptors, causing initially stimulation and at higher doses blockade of the ganglionic receptors similar to the effects of curare. Pregnant animals that consume the seeds or leaves over a period of time may experience the teratogenic effects of the alkaloid anagyrine as seen in cattle consuming lupines in early pregnancy (arthrogryposis, cleft palate).

Similar quinolizidine alkaloids are found in members of the genus *Cytisus* (Scotch broom, broom). These plants are quite commonly grown for their foliage and profusion of yellow or white pea-like flowers. The potential for similar toxicity to that occurring with *Laburnum* species is therefore present, especially if the seeds are eaten in quantity.

Risk Assessment
Laburnum species, and especially their hybrid *L. watereri*, are commonly grown in mild temperate climates for the striking display of pendulant inflorescences of yellow flowers. The production of numerous seed pods and seeds increases the chances of children or household pets ingesting the seeds.

Clinical Signs
The most prevalent signs of poisoning in dogs are those of vomiting and abdominal pain, and to a lesser extent weakness, depression, ataxia, and tachycardia.[3,4] The signs are usually short-lived, and recovery is common. In severe cases where large quantities of the seeds are consumed, myocardial degeneration may lead to death.

Treatment is rarely necessary, and when necessary should include the oral administration of activated charcoal and other supportive therapy to counter the clinical effects of vomiting and abdominal pain.

References

1. Habermehl GG: Poisonous Plants of Europe. In Poisonous Plants: Proceedings of the third International symposium. James LF, Keeler RF, Bailey EM, Cheeke PR, Hegarty MP eds. Iowa State University Press, Ames pp 74-83, 1977.
2. Keeler RF, Baker DC: Myopathy in cattle induced by alkaloid extracts from *Thermopsis montana, Laburnum anagyroides,* and a *Lupinus* sp. J Comp Pathol 103: 169-182, 1990.
3. Clarke ML. Clarke EG. King T. Fatal laburnum poisoning in a dog. Vet Rec 88: 199-200, 1971.
4. Leyland A. Laburnum *(Cytisus laburnum)* poisoning in two dogs. Vet Rec 109: 287, 1981.

Lantana

Family: Verbenaceae

Common Names

Lantana, shrub verbena, red sage, wild sage, corona del sol, cinco negritos

Plant Description

A genus of about 150 species, *Lantana* are evergreen perennials from tropical areas of the Americas and southern Africa. As woody shrubs with 4-sided stems, some spiny, growing 2-3m in height, *Lantana* species have leaves that are simple, toothed, opposite or in whorls. The leaves have a strong smell when crushed. Inflorescences are terminal, rounded with individual flowers being tightly grouped in the heads. Flowers are showy, 5 lobed, salverform, with varying colors of white, orange, yellow, red, or purple depending on the species. The petal colors change as the newest flowers located in the center of the flower head mature, turning from yellow to red or purple, white to pink, red to darker shades of red (Figures 241 and 242). The fruits are round drupes turning glossy black when ripe, and containing 2 seeds (Figure 243).

Figure 241 *Lantana camara*

Figure 242 *Lantana* flower

The 2 most common species of toxicologic significance are: *Lantana camara*, and *L. montevidensis* – creeping or trailing lantana.

Figure 243 *Lantana camara* flowers and fruits

Toxic Principle and Mechanism of Action

The pentacyclic triterpenoid lantadenes A, B, and C are the primary cumulative hepatotoxins in *Lantana*, and are present in all parts of the plant.[1-4] However, not all species of *Lantana* are toxic. The lantadenes are absorbed from the intestinal tract and are biotransformed in the liver causing damage to the bile canuliculi with blockage of bile flow and cholestasis.[5-7] As a result the excretion of phylloerythrin is impaired, which causes a secondary photosensitivity in nonpigmented skin exposed to sunlight. Lantadenes are also toxic to the kidneys and cause gastroenteritis.[6] *Lantana* causes poisoning in a wide variety of animals including cattle, sheep, goats, horses, dogs, guinea pigs, and kangaroos.[8-12]

Risk Assessment

Lantana is a common garden plant in tropical areas, and is also grown as a potted house plant in temperate areas. In tropical areas, the plant can become a noxious weed, and is a source of poisoning to livestock. It is rarely associated with poisoning of household pets, but dogs and children are occasionally poisoned by eating the leaves and fruits.[12,13]

The green fruits are apparently attractive and palatable, but quite toxic to children causing vomiting, weakness, lethargy, dilated pupils, and unconsciousness.[13] The blue-black ripe fruits are not toxic, but partially ripe fruit may retain some toxicity.

Clinical Signs

Vomiting, diarrhea, labored respiration, and weakness occur following the ingestion of the unripe berries in children.[13] Similar signs, including liver failure, have been reported in dogs.[12] Livestock are most frequently poisoned by *Lantana*, and initially affected animals become anorexic, depressed, and constipated.[10,11] Within a few days icterus develops and after a few days to a few weeks the animals that have nonpigmented skin show signs of secondary photosensitization. Affected white skin becomes reddened, swollen, and dies resulting in dried leathery skin that sloughs off. Affected animals become hypersensitive to sunlight and seek shade. Weight loss, diarrhea, and eventually death occur as a result of liver and kidney failure.[7]

Treatment of animals showing signs of liver failure is not successful owing to the severity of the liver degeneration. Livestock that have eaten quantities of *Lantana* very recently can be given bentonite or activated charcoal orally to reduce absorbtion of the *Lantadine* toxins.[14]

On postmortem examination, the gall bladder can be distended and the liver yellow in color due to billiary statsis. The kidneys are usually swollen and perirenal edema can be present. Histologically the liver shows periportal degeneration, fibrosis, and biliary hyperplasia.[7]

References

1. Sharma OP, Vaid J, Pattabhi V, Bhutani KK: Biological action of lantadene C, a new hepatotoxicant from *Lantana camara* var. aculeata. J Biochem Toxicol 7: 737-739, 1992.
2. Sharma OP, Makkar HP, Dawra RK: A review of the noxious plant *Lantana camara.* Toxicon 26: 975-987, 1988.
3. Sharma OP: Review of the biochemical effects of *Lantana camara* toxicity. Vet Human Toxicol 26: 488-493, 1984.
4. Ghisalberti EL: *Lantana camara* L. (Verbenaceae). Fitoterapia 71: 467-486, 2000.
5. Sharma S, Sharma OP, Singh B, Bhat TK. Biotransformation of lantadenes, the pentacylclic triterpenoid hepatoxins of lantana plant in guinea pigs. Toxicon 38: 1191-1202, 2000.
6. Pass MA, Pollit S, Goosem MW, McSweeney CS: The pathogenisis of lantana poisoning in plant toxicology. In Proceedings of the Australian-USA Poisonous Plants Symposium, Animal Research Institute, Yeerongopilly, Brisbane, Australia. 487-494, 1985.
7. Seawright AA, Allen JG: Pathology of liver and kidney in lantana poisoning of cattle. Aust Vet J 48: 323-321, 1972.
8. Johnson JH, Jensen JM: Hepatotoxicity and secondary photosensitization in a red kangaroo *(Megaleia rufus)* due to ingestion of *Lantana camara.* J Zoo Wildl Med 29: 203-207, 1998.
9. Ide A. Tutt CL: Acute *Lantana camara* poisoning in a Boer goat kid. J South African Vet Assoc 69(1): 30-2, 1998
10. Ganai GN, Jha GJ: Immunosuppression due to chronic *Lantana camara*, L. toxicity in sheep. Indian J Expl Biol 29(8): 762-6, 1991.
11. Fourie N, an der Lugt JJ, Newsholme SJ, Nel PW: Acute *Lantana camara* toxicity in cattle. J South African Vet Assoc 58: 173-178, 1987.
12. Morton JF: Lantana or red sage (*Lantana camara* L.,(Verbenaceae)), notorious weed and popular garden flower: some cases of poisoning in Florida. Econ Bot 48: 259-270, 1994.
13. Wolfson SL, Solomons TWG: Poisoning by fruit of *Lantana camara.* Am J Dis Child 107: 173-176, 1964.
14. McKenzie RA: Bentonite as therapy for *Lantana camara* poisoning of cattle. Austral Vet J 68: 146-148, 1991.

Family: Fabaceae

Figure 244 *Lathyrus latifoluis*

Figure 245 *Lathyrus latifoluis* flowers

Common Names

Sweet pea *(Lathyrus odoratus)*, perennial or everlasting pea *(L. latifolius)*

Plant Description

Consisting of about 150 species, the genus *Lathyrus* is closely related to the genera *Pisum* (garden peas) and *Vicia* (vetches). Common to the northern temperate climates, *Lathyrus* species are annual or perennial, typically climbing or trailing vines, with characteristic tendrils and winged stems. Some species have a taproot, while others have rhizomatous roots. Leaves are pinnately compound, with 2-14 oblong-lanceolate leaflets, but with tendrils in place of a terminal leaflet. Inflorescences are axillary racemes or single flowers. Flowers have 5 fused sepals, and 5 petals, comprising a prominent standard or banner petal, wings, and a keel. Flower colors come in white, pink, red, purple, blue or yellow (Figures 244 and 245). The fruits are typical legume pods that split open when ripe to release numerous round, brown seeds.

Toxic Principle and Mechanism of Action

Amino acids and aminonitriles present especially in the seeds but also in the green plant are responsible for a variety of neurologic, musculoskeletal, and blood vessel abnormalities collectively referred to as lathyrism.[1] The amino acid beta-aminoproprionitile (BAPN) is one of the principle toxins as it interferes with lysyl oxidase, a critical enzyme associated with cross-linking of collagen, and therefore its strength and stability. Consequently, humans and animals eating quantities of the seeds over a period of weeks develop defective cartilage that manifests as bone pain, abnormal gait, and lameness. Defects in the collagen of blood vessels can result in vessel rupture and acute death. Monogastric animals such as humans and horses, and to a lesser extent ruminants, develop neurolathyrism, a disease associated with another compound, beta-N-oxalyl-amino-L-alanine, that when present at high enough levels causes excessive activity of the excitatory receptors in neurons that mimics glutamate toxicity. This can lead to permanent neuron degeneration resulting in muscle weakness progressing to paralysis.

Risk Assessment

Both the common sweet pea *(Lathyrus odoratus)*, and the perennial or everlasting pea *(L. latifolius)* contain toxic amino acids and nitriles and have the potential for causing toxicity in humans, horses, sheep, cattle, pigs, poultry, monkeys, and elephants that eat the seeds.[2] As there are numerous species of *Lathyrus* that contain these neurotoxins or lathyrogens, it is reasonable to assume all members of the genus are toxic until proven otherwise.[2] The seeds and vines of sweet peas should not be fed to animals or birds.

Clinical Signs

Neurolathyrism in horses is characterized by muscle weakness, stiff gait, incoordination, difficulty in getting-up after laying down. The disease is progressive as long as the animal continues to eat the *Lathyrus* seeds or hay, and will result in permanent neuromuscular degeneration and paraplegia. Laryngeal paralysis, with distinctive "roaring" respiratory sounds is not uncommon in the horse.[3] Deaths may be acute when large blood vessels rupture due to angiolathyrism.

Signs of lathyrism in ruminants are associated with muscle weakness, difficulty in walking, muscle tremors, and recumbency.[3] Sheep may show more signs of neurological disease such as circling, head pressing, abdominal pain, and seizures. Animals generally recover if they have not consumed the *Lathyrus* seeds and hay for too long a period.

Chickens, ducks, turkeys, and other birds eating the seeds may develop a variety of neurologic signs including twisted or distorted necks, seizures, rupture of arteries, large deformed eggs, and decreased egg production.[4,5]

Lathyrism is a serious problem in human populations in some parts of the world where the seeds of *Lathyrus* species are consumed in quantity in times of drought and famine.[6] Initially people develop weakness of the legs, and difficulty in walking. Spacticity of the muscles often leads to the inability to walk and total incapacitation of the person.

Treatment is generally aimed at supportive care and removing annual perennial peas from the diet. Mild cases will recover, but the neurological changes tend to be permanent in more chronically poisoned animals.

References

1. Spencer PS. Schaumburg HH. Lathyrism: a neurotoxic disease. Neurobehavioral Toxicol Teratology 5: 625-629, 1983.
2. Bell EA: Aminonitriles and amino acids not derived from proteins. In Toxicants Occurring Naturally in Foods 2nd ed. National Research Council, National Academy of Sciences, Washington, DC. 153-169, 1973.
3. Burrows GE, Tyrl RJ: Toxic Plants of North America. Iowa State University Press, Ames. 564-572, 2001.
4. Chowdhury CD: Effects of low and high dietary levels of beta-aminoproprion-itrile (BAPN) on the performance of laying hens. J Sci Food Agric 52: 315-320, 1990.
5. Terpin T. Roach MR. A biophysical and histological analysis of factors that lead to aortic rupture in normal and lathyritic turkeys. Can Physiol Pharmacol 65: 395-400, 1987,
6. Haimanot RT et al: The epidemiology of lathyrism in north and central Ethiopia.[erratum appears in Ethiop Med J 1993, 31:155-156]. Ethiopian Medical J 31: 15-24, 1993.

Leucothoe
Family: Ericaceae

Figure 246 *Leucothoe fontanesiana*

Common Name
Fetterbush

Plant Description
Comprising a genus of about 50 species, *Leucothoe* are native to North and South America, Asia, and Madagascar. Deciduous or evergreen woody shrubs, with arching branches, alternate, glossy green, elliptic to lanceolate leaves with serrated edges. Infloresences are axillary racemes of tubular, bell-shaped, white flowers with 5 fused sepals and 5 fused petals (Figure 246). Fruits are capsules.

Toxic Principle and Mechanism of Action

All species of the family Ericaceae contain varying quantities of toxic diterpenoids collectively known as grayanotoxins I and II (formerly andromedotoxin, rhodotoxin, and acetylandromedol).[1] As many as 18 grayanotoxins (I – XVIII) have been identified, the greatest number being found in the *Leucothoe* species (fetterbush).[2,3] Tannins and other compounds are also present in varying amounts. All parts of the fetterbrush including the flowers and the nectar are toxic, although there may be considerable variation between species and even amongst plants of the same species depending on the growing conditions.

Grayanotoxins act to increase sodium channel permeability of cells by opening the channels to sodium, which enters the cells in exchange for calcium ions, thus rendering the channels slow to close so that the cell remains depolarized.[4,5] Other neurologic mechanisms may also involve a cholinergic response seen clinically as bradycardia and excessive salivation.[6] The cardiac effects can range from bradycardia, sinus arrest, and arrhythmias.

Other members of the Ericaceae (Heath family) that contain grayanotoxins include:

Andromeda polifolia	Andromeda, bog rosemary
Kalmia spp.	Laurels
Ledum spp.	Labrador tea
Lyonia spp.	Maleberry
Menziesia spp.	Rusty menziesia
Pieris spp.	Pieris, Japanese pieris
Rhododendron spp.	Rhododendron, azalea

Risk Assessment
Commonly grown as showy garden plants, especially in wooded areas, fetter bush is seldom associated with poisoning. Like rhododendrons and other members of the Heath family, the plants are toxic to animals especially in winter time when the evergreen leaves are an attraction. Honey made by bees feeding on the nectar of the flowers has been known to be toxic to people who eat the "mad-honey".[7]

Clinical Signs
Excessive salivation, increased nasal secretions, vomiting, abdominal pain, bloat, and irregular respirations develop several hours later as in rhododendron poisoning.[8-10] Projectile vomiting may be noticeable. Hypotension, tachycardia, and respiratory depression may also develop. Weakness, partial blindness, and seizures have been reported in severe intoxications. Neurologic signs may persist for several days before the animal recovers. Weight loss may be notable.

Treatment is primarily directed at relief of the more severe clinical signs. Activated charcoal given orally is helpful if given shortly after the plant is consumed. Atropine is useful in countering the cardiovascular effects

References
1. Kakisawa H, Kozima T, Yanai M, Nakanishi K: Stereochemistry of grayanotoxins. Tetrahedron 21: 3091-3104, 1965.
2. Sakikabara J, Shirai N, Kaiya T, Nakata H: Grayanotoxin-XVIII and its grayanoside B, a new A-Nor-B-Homo-Ent-Kaurine and its glucoside from *Leucothoe grayana*. Phytochemistry 18: 135-137, 1979.
3. Sakikabara J, Shirai N, Kaiya T: Diterpene glycosides from *Pieris japonica*. Phytochemistry 20: 1744-1745, 1981.
4. Narahashi T, Seyama I: Mechanism of nerve membrane depolarization caused by grayanotoxin I. J Physiol 242: 471-487, 1974.
5. Seyama I, Narahashi T: Modulation of sodium channels of squid nerve membranes by grayanotoxin I. J Pharmacol Exp Ther 219: 614-624, 1981.
6. Onat F, Yegan BC, Lawrence R, Oktay A, Oktay S: Site of action of grayanotoxins in mad honey in rats. J Appl Toxicol 11: 199-201, 1001.
7. Krochmal C: Poison honeys. Am Bee J 134: 549-550, 1994.

Family: Asteraceae

Figure 247 Ligularia stenocephala

Common Name
Ligularia

Plant Description
There are approximately 150 species of *Ligularia* native to the temperate area of Asia and Europe, and are closely related to the genus *Senecio (Packera)*. Many are large-leafed, clump forming perennials preferring moist, shady areas. Infloresences are racemes on erect stems, purple in some species *(L. stenocephala)*. Flowers are numerous, yellow to orange, daisy-like, and produced on tall spires as much as 1.5 m in height (Figure 247).

Toxic Principle and Mechanism of Action
Some species of *Ligularia* have been shown to contain pyrrolizidine alkaloids (PA) with similar toxicity to those found in species of *Senecio*.[1,2,3] These alkaloids are converted by the liver into toxic pyrroles that inhibit cellular protein synthesis and cell mitosis.[3-5] Liver necrosis, degeneration, and fibrosis with biliary hyperplasia characterize the toxic effects of the pyrrolizidine alkaloids. There is considerable variation in the toxicity of the alkaloids depending upon the species of plant.

Risk Assessment
Poisoning of domestic animals by *Ligularia* species has not been reported although yaks have been suspected of being poisoned.[2] However, the presence of pyrrolizidine alkaloids in *Ligularia* species makes these plants potentially hazardous to animals.

Clinical Signs
Animals that consume PA containing plants over a period of time develop signs related to liver failure. Horses and cattle are generally the most severely affected, while sheep and goats are quite resistant to PA toxicity. Weight loss, icterus, diarrhea, photosensitization, and neurologic signs related to hepatic encephalopathy are typical of liver failure. Serum liver enzymes are generally elevated significantly. Confirmation of PA toxicity can be made by a liver a biopsy showing the triad of histologic changes characteristic of PA poisoning, namely liver megalocytosis, fibrosis, and biliary hyperplasia.

References
1. Ji LL, Zhao XG, Chen L, Zhang M, Wang ZT: Pyrrolizidine alkaloid clivorine inhibits human normal liver L-02 cells growth and activates p38 mitogen-activated protein kinase in L-02 cells. Toxicon 40: 1685-1690, 2002.
2. Winter H, et al: Pyrrolizidine alkaloid poisoning of yaks: identification of the plants involved. Veterinary Record 134: 135-139, 1994.
3. Mattocks AR: Chemistry and toxicology of pyrrolizidine alkaloids. Academic Press, Orlando, Florida, 1986.

Ligustrum
Family: Oleaceae

Common Names
Japanese privet *(L. japonicum)*, common privet *(L. vulgare)*, Chinese privet *(L. sinense)*, California privet *(L. ovalifolium)*, and amur privet *(L. amurense)* are some of the more commonly encountered species in North America.

Plant Description
A genus of about 50 species of shrubs and small trees that are evergreen or deciduous in their native habitats in Europe, North Africa, and Australasia. Leaves are glossy green, opposite, elliptical to ovate, with entire edges. Inflorescences are terminal panicles of small, white, scented, tubular, 4-lobed flowers (Figure 248). Fruits are purple-black berries that often persist after the leaves drop (Figure 249).

Figure 248 *Ligustrum vulgare* flowers

Toxic Principle and Mechanism of Action
A variety of compounds are present in the leaves and fruits, but the specific toxins responsible for the irritant effects on the gastrointestinal system are terpenoid glycosides of oleanolic acid.[1] All parts of the plant are toxic.

Risk Assessment
Privets are common garden and landscape plants grown for their glossy green foliage

Figure 249 *Ligustrum* berries

and displays of scented white flowers. Some species make good hedges. Planting privets around animal enclosures should be discouraged. The black berries that persist after the leaves have dropped off are also tempting to children.

Clinical Signs
Privet poisoning has been suspected in people, cattle, sheep, and horses.[2] Vomiting, colic, and diarrhea are common signs of privet poisoning. In severe cases, ataxia, recumbency, increased heart and respiratory rates, and death may result where quantities of the leaves and berries are consumed.[3]

References
1. Connolly JD, Hill RA: Dictionary of Terpenoids. Vol 1. Mono- and sesquiter-penoids. Chapman & Hall. London, 1991a.
2. Burrows GE, Tyrl RJ. Toxic Plants of North America. Iowa State University Press, Ames. 834-837, 2001.
3. Kerr LA, Kelch WJ: Fatal privet *(Ligustrum amurase)* toxicosis in Tennessee cows Vet Hum Toxicol 41: 391-392, 1991.

Lilium

Family: Liliaceae

Figure 250 *Lilium superbum* (Turkscap lily)

Common Names
Easter lily, tiger lily, Asiatic lily. (Day lilies belong to the genus *Hemerocallis*)

Plant Description
The genus *Lilium* has approximately 100 species that are indigenous to Europe and Asia and North America. All species grow from bulbs consisting of overlapping fleshy scales that do not encircle the bulb as in the onion-type bulb. Leaves are arranged in spirals or whorls on the erect stems, and vary from grass-like, linear, to lanceolate. The inflorescences consist of solitary flowers, racemes or umbels, with flowers being held erect, horizontal or pendent, and are generally large showy and cup or funnel-shaped. Flower colors include white, yellow, orange, red, or maroon with frequent spotting on the inner surfaces of the petals (Figures 250 and 251). Numerous hybrids have been developed and are widely available commercially. The most common is the Easter lily *(Lilium longiflorum)* (Figure 252).

Toxic Principle and Mechanism of Action
The common Easter lily, the tiger lily, Asiatic or Japanese lily, and the numerous *Lilium* hybrids and day lilies (Hemerocallis spp.) are toxic to cats causing kidney failure that can prove to be fatal.[1-5] The toxin responsible for the nephrotoxicity of lilies has not been identified. All parts of the plants are poisonous to cats, but especially the flowers. Deaths have been reported in cats after ingestion of only two leaves. Dogs, rats, and rabbits were not affected after they were fed high doses of Easter lily experimentally.[3]

Nephrotoxicity is known to occur in Britain, Norway, and Japan following the ingestion of the lily known as bog asphodel *(Narthecium ossifragum* and *N. asiaticum)*.[6,7] Consumption of the flowers in particular causes renal tubular necrosis in cattle and sheep, suggesting that *Lilium* species contain similar toxins.[4,5]

Risk Assessment
Easter lilies, day lilies, Asiatic lilies and their hybrids are very popular house and garden plants and are a significant risk to indoor cats that are prone to chewing on plants. Easter lilies are one of the most toxic household plants for cats and all *Lilium* species should be considered toxic unless proven otherwise.

Clinical Signs

Within 3 hours of consuming 1-2 leaves or flower petals of the lily, cats start to vomit and salivate excessively. The cats become depressed, anorexic, with the initial vomiting and salivation tending to subside after 4-6 hours. Approximately 24 hours later, proteinuria, urinary casts, isosthenuria, polyuria, and dehydration develop. Vomiting may recur at this stage. A disproportionate increase in serum creatinine as compared to blood urea nitrogen is a significant indicator of lily poisoning. As the cat develops, renal failure, anuria, progressive weakness, recumbency, and death occur.[1-4]

Figure 251 *Lilium* hybrid

On postmortem examination the kidneys are swollen and perirenal edema may be evident. Mild to severe pulmonary and hepatic congestion is common. Renal tubular necrosis is the most prominent histopathologic lesion. Granular and hyaline casts are common in the collecting ducts.[3]

Treatment

Cats suspected of eating lilies should be seen by a veterinarian as soon as possible after the plant was consumed. Gastrointestinal decontamination and fluid therapy is essential for preventing the nephrotoxicity. Vomiting should be induced if the plant has been consumed within the last 1-2 hours. Activated

Figure 252 *Lilium longiflorum* (Easter lily)

charcoal with a cathartic should be given to decontaminate the gastrointestinal tract. Fluid therapy should be initiated to maintain renal function and prevent anuria, and should be continued for at least 24 hours. Once the cat has developed anuric renal failure, peritoneal dialysis or hemodialysis will be necessary to try and save the animal.

References

1. Hall JO: Nephrotoxicity of Easter lily *(Lilium longiflorum)* when ingested by the cat. Proc Am Coll Vet Intern Med Forum 10: 840, 1992.
2. Carson TL, Sanderson TP, Halbur PG: Acute nephrotoxicity process in cats falling ingestion of Lilium *(Lilium* spp.). Am Assoc Vet Lab Diagn 37: 43, 1994.
3. Hall JO: Lily. In clinical veterinary toxicology. Plumlee KH ed. Moseby, St. Louis, Missouri pp. 433-435, 2003.

4. Langston CE: Acute renal failure caused lily ingestion in 6 cats. J Am Med Assoc 220: 49-52, 2002.

5. Rumbeiha WK, Francis JA, Fitzgerald SD et al: A comprehensive study of Easter lily poisoning in cats. J. Vet Diagn Invest 16:527-541, 2004

6. Flaoyen A, et al: Nephrotoxicity of *Narthecium ossifragum* in cattle in Norway. Vet Rec 137: 259-263, 1995.

7. Flaoyen A: *Narthecium ossifragum* associated nephrotoxicity in ruminants. In toxic plants and other natural toxicants. Garland T, Barr AC. eds. CAB International, New York pp 573-576, 1998.

Lobelia

Family: Campanulaceae

Figure 253 *Lobelia cardinalis*

Common Names

Cardinal flower
The following 4 species have toxicologic significance.
L. berlandieri - Berlandier lobelia
L. cardinalis - cardinal flower, Indian pink
L. inflata - Indian tobacco, emetic weed, eye bright
L. siphilitica - blue lobelia, great lobelia, Louisiana lobelia

Plant Description

Comprising a genus of about 370 species of annual and perennial herbs or shrubs, *Lobelia* are native to tropical and temperate areas of North, Central, and South America. Plants are erect, 60-150cm tall, branched or not branched, with leaves that are sessile or petiolate, lanceolate to elliptic, margins entire or serrate. Flowers are produced in terminal racemes or singly in leaf axils. The 2 upper lobes are smaller than the 3 lower lobes. Flower color range includes white, blue, lavender, and scarlet (Figures 253-255). Fruits are capsules with many seeds.

Toxic Principle and Mechanism of Action

A variety of pyridine alkaloids with nicotine-like properties are found in all parts of the plant, and especially in the mature plant. Three major groups of alkaloid are present including lobelines, lelobines, and lobinines.[2] The alkaloid content of the leaves range from 0.08 - 0.58% depending on the species, with highest concentrations in *L. inflate* and *L. sipilitica*.[3] The lobeline alkaloids are potent stimulants of nicotinic ganglia at low doses and cause paralysis at high doses. Heart rates are reduced, and cardiac irregularities may develop. Lobeline has stimulatory effects on the central nervous system but to a lesser degree than nicotine. Most poisoning from *Lobelia* has been reported in cattle, sheep, and goats.[4]

Risk Assessment

Popular as garden plants for their bright colorful flowers, lobelias do well as plants for water or bog gardens as they prefer wet growing conditions with the exception of *L. berlandieri* which is native to the drier conditions of Texas and Mexico. It is responsible for most cases of poisoning of cattle. Poisoning of animals from garden grown Lobelias is infrequent, but care should be taken not to feed garden clippings containing lobelia plants to livestock. Poisoning in humans occurs when *Lobelia* preparations are used as home remedies.

Figure 254 *Lobelia* cultivar

Clinical Signs

In cattle the signs of *L. berlandieri* poisoning include depression, diarrhea, excessive nasal discharge, pupillary dilation, labored respiration, and incoordination.[4] In severe cases, death appears to be due to cardiopulmonary failure. At post mortem examination, there is evidence of congestion of the brain, hyperinflated lungs, hemorrhaging on organs, and ascites. Ulceration throughout the small intestine, necrosis of the mucosa of the intestines, the liver, and kidneys may be evident.

Figure 255 *Lobelia siphilitica*

References

1. Burrows GE, Tyrl RJ. Toxic Plants of North America. Iowa State University Press, Ames. 312-315, 2001.
2. Marion L: The pyridine alkaoids. In the Alkaloids, vol 6. Manske RHF ed. Academic Press, New York. 123-144, 1960.
3. Krochmal A, Wilken L, Chien M: Lobeline content from four Appalachian lobelias. Lloydia 35: 303-304, 1972.
4. Dollahite JW, Allen TJ: Poisoning of cattle, sheep, and goats with *Lobelia* and *Centaurium* species. Southwest Vet 182: 126-130, 1969.

Family: Caprifoliaceae

Figure 256 *Lonicera sempervirens*

Figure 257 *Lonicera Heckrottii*

Common Names

Honeysuckle, woodbine, twin berry
The more common species are *L. involucrata* (black twin berry), *L. Japonica* (Japanese honeysuckle), *L. peirclymenum* (common honeysuckle), and *L. sempervirens* (trumpet or coral honeysuckle).

Plant Description

Native to the temperate and subtropical regions of the Northern Hemisphere, there are some 180 species *Lonicera*, with about 36 species being native or introduced to North America. Deciduous or evergreen shrubs or woody vines, with simple, opposite, green leaves and inflorescences of 2-6 flowers produced in leaf axils or terminally. Flowers are tubular or with the upper lip having 4 lobes and the lower lip unlobed. Petals are white, yellow, orange, red, or pink. Pistils are prominent (Figures 256-258). Fruits are red, blue, or black round berries with few seeds (Figure 259).

Toxic Principle and Mechanism of Action

A variety of potentially toxic compounds including triterpenoid saponins similar to those in common English ivy (*Hedera* spp.) have been found in some *Lonicera* species.[1] These compounds are irritants and may cause gastrointestinal irritation. Not all *Lonicera* species are toxic.

Risk Assessment

Grown for their attractive perfumed flowers and ability to grow rapidly, honeysuckles are commonly found in gardens growing along fences, trellises, or as hedges. Poisoning of animals is rare, but children may be affected after eating the red, blue, or black berries especially of the European species.[2,3]

Clinical Signs

Vomiting and diarrhea can be anticipated if quantities of the berries are consumed. Treatment is generally symptomatic.

References

1. Son KH, Jung KY, Chang HW, Kim HP, Kang SS: Triterpenoid saponins from the aerial parts of *Lonicera japonica*. Phytochem 35: 1005-1008, 1994.
2. Lamminpaa A, Kinos M: Plant poisonings in children. Hum Expl Toxicol 15: 245-249, 1996.
3. Lampe KF, McCann MA: AMA Handbook of Poisonous and Injurious Plants. Am Med Assoc, Chicago, Illinois pp 111-112, 1985.

Figure 258 *Lonicera japonica*

Figure 259 *Lonicera spp.* fruits

Lyonia

Family: Ericaceae (Heath family)

Common Names

Stagger bush, maleberry, male blueberry Poisonous species include *L. ligustrina* and *L. mariana*.

Plant Description

Comprising a genus of 30-35 species native to North America, Eastern Asia, Mexico, and the West Indies, *Lyonia* are small to large, deciduous or evergreen woody, branching shrubs. The leaves are alternate, glossy green, elliptic to oblong.

Figure 260 *Lyonia mariana*

Inflorescences are axillary racemes of bell-shaped or cylindrical white flowers with arching branches, alternate, glossy green, elliptical to lanceolate leaves with serrated edges. Infloresences are axillary racemes of tubular, bell-shaped, white flowers with 5 fused sepals and 5 fused petals (Figure 260). Fruits are capsules.

Toxic Principle and Mechanism of Action

All species of the family Ericaceae contain varying quantities of toxic diterpenoids collectively known as grayanotoxins I and II (formerly andromedotoxin, rhodotoxin, and acetylandromedol).[1] As many as 18 grayanotoxins (I – XVIII) have been identified, the greatest number being found in the *Leucothoe* species (fetter bush).[2,3] Tannins and other compounds are also present in varying amounts. All parts of the rhododendrons including the flowers and the nectar are toxic, although there may be considerable variation between species and even amongst plants of the same species depending on the growing conditions. Grayanotoxins act to increase sodium channel permeability of cells by opening the channels to sodium, which enters the cells in exchange for calcium ions, thus rendering the channels slow to close so that the cell remains depolarized.[4,5] Other neurologic mechanisms may also involve a cholinergic response seen clinically as bradycardia and excessive salivation.[6] The cardiac effects can range from bradycardia, sinus arrest, and arrhythmias.

Other members of the Ericaceae that contain grayanotoxins include:

Andromeda polifolia	Andromeda, bog rosemary
Kalmia spp.	Laurels
Ledum spp.	Labrador tea
Leucothoe spp.	Fetterbush
Pieris spp.	Pieris, Japanese pieris
Rhododendron spp.	Rhododendron, azalea

Risk Assessment

Occasionally grown as woodland garden plants, *Lyonia* are potentially poisonous to people and animals like other members of the Heath family (Ericaceae)

Clinical Signs

Excessive salivation, increased nasal secretions, vomiting, abdominal pain, bloat, and irregular respirations develop after several hours as for rhododendron poisoning.[8-10] Projectile vomiting may be noticeable. Hypotension, tachycardia, and respiratory depression may also develop. Weakness, partial blindness, and seizures have been reported in severe intoxications. Neurologic signs may persist for several days before the animal recovers. Weight loss may be notable.

Treatment is primarily directed at relief of the more severe clinical signs. Activated charcoal given orally is helpful if given shortly after the plant is consumed. Atropine is useful in countering the cardiovascular effects

References

1. Kakisawa H, Kozima T, Yanai M, Nakanishi K: Stereochemistry of grayanotoxins. Tetrahedron 21: 3091-3104, 1965.

2. Sakikabara J, Shirai N, Kaiya T, Nakata H: Grayanotoxin-XVIII and its grayanoside B, a new A-Nor-B-Homo-Ent-Kaurine and its glucoside from *Leucothoe grayana*. Phytochemistry 18: 135-137, 1979.
3. Sakikabara J, Shirai N, Kaiya T: Diterpene glycosides from *Pieris japonica*. Phytochemistry 20: 1744-1745, 1981.
4. Narahashi T, Seyama I: Mechanism of nerve membrane depolarization caused by grayanotoxin I. J Physiol 242: 471-487, 1974.
5. Seyama I, Narahashi T: Modulation of sodium channels of squid nerve membranes by grayanotoxin I. J Pharmacol Exp Ther 219: 614-624, 1981.
6. Onat F, Yegan BC, Lawrence R, Oktay A, Oktay S: Site of action of grayanotoxins in mad honey in rats. J Appl Toxicol 11: 199-201, 1991.
7. Krochmal C: Poison honeys. Am Bee J 134: 549-550, 1994.

Macadamia

Family: Protaceae

Common Names
Macadamia nut, Queensland nut *(Macadamia integrifolia)*, and rough shell macadamia or bopple nut *(M. tetraphylla)*.

Plant Description
A relatively small genus of 11 species of medium sized trees native to Australia, Caledonia, and Indonesia that are commonly grown commercially in Hawaii and California. As evergreen rain forest trees, they have evergreen, leathery, lanceolate, serrated or entire, leaves arranged in whorls of 3-4 on the twigs. Inflorescences consist of many small white to pinkish flowers crowded on pendent cylindrical spikes. The fruits are nuts which take up to 9 months to mature. The edible nuts are encased in a tough endocarp or shell that must be cracked to access the fruit (Figures 261 and 262).

Figure 261 *Macadamia integrifolia*

Figure 262 *Macadamia* nuts (shelled)

Toxic Principle and Mechanism of Action
The toxin in macadamia nuts responsible for causing poisoning in dogs has not been identified. The mean dose of the nuts that will cause poisoning is 11.7 mg/kg body weight, with signs of intoxication appearing in less than 12 hours of consuming the nuts.[1] Dogs experimentally dosed with 20 gm macadamia nuts/kg body weight orally developed signs of poisoning within 12 hours, and exhibited weakness, inability

to rise, central nervous system depression, vomiting, and hyperthermia with rectal temperatures up to 45.5°C.[1] Recovery occurs in 24 hours.
Macadamia nuts contain significant quantities of palmitoleic acid, oleic acid, linoleic acid, and linolenic acid.[2] In addition there are significant amounts of alpha-tocopherol and various other sterols. Similar profiles of monounsaturated fatty acids are found in almonds, peanuts, hazel nuts, and walnuts.[2]

Risk Assessment
Dogs are unlikely to be affected by the nuts that may fall from the tree as the nuts are too tough for dogs to break open. Most poisonings occur when a dog is fed the shelled nuts or eats cookies made with macadamia nuts.

Clinical Signs
Within 12 hours of eating the nuts, dogs develop weakness, ataxia, depression, vomiting, muscle tremors, stiffness, lameness, inability to rise, and hyperthermia.[1] Elevated serum alkaline phosphatase, triglyceride and lipase levels may be elevated, returning to normal in 48 hours.[1] Treatment is seldom necessary as dogs recover in about 24 hours.

References
1. Hansen SR, Buck WB, Meerdink G, Khan SA: Weakness, tremors, and depression associated with macadamia nuts in dogs. Vet Human Toxicol 42: 18-21, 2000.
2. Maguire LS, O'Sullivan SM, Galvin K, O'Connor TP, O'Brien NM: Fatty acid profile, tocopherol, squalene and phytosterol content of walnuts, almonds, peanuts, hazelnuts and the macadamia nut. Internat J Food Sci Nutrition 55: 171-178, 2004.

Melaleuca
Family: Myrtaceae

Common Names
Parebark, melaleuca,
Tea tree *(Melaleuca alternifolia)*, broad leafed paperbark, punk tree
(M. quinquenervia), the latter species being a noxious weed in Florida.
Melaleuca are closely related to the genus *Eucalyptus.*

Plant Description
A large genus of 150 shrubs and trees, *Melaleuca* are indigenous principally to Australia and Southeast Asia. Some species have a papery bark that peels off in sheets. Leaves are generally simple, leathery, and either flat or cylindrical.

Figure 263 *Melaleuca quinquenervia*

Inflorescences are profuse, bottlebrush-like spikes, with flowers having showy stamens and colors of white, pink, purple, or red (Figures 263 and 264).

Toxic Principle and Mechanism of Action

Melaleuca alternifolia contains an essential, pungent smelling, colorless or light yellow oil similar to Eucalyptus oil. Toxicity is due to the presence of cyclic hydrocarbon terpenes, sesquiterpenes, and various oils that are readily absorbed through the skin and mucous membranes.[1] The mechanism of toxicity

Figure 264 *Melaleuca quinquenervia* flowers

has not been determined. Melaleuca oil has antibacterial and antifungal properties and has been used topically on dogs and cats to treat skin infections and repel fleas. The oil is also toxic to people as it is readily absorbed through the skin or if ingested.[2-4] Skin absorption is increased if the melaleuca oil is applied with organic solvents such as alcohol or dimethylsulfoxide (DMSO).

Melaleuca have not been associated with cyanide poisoning that has been encountered in sheep, cattle, and goats eating the leaves from recently felled trees of *Eucalyptus cladocalyx* (sugar gum) and *E. viminalis* (manna gum).[5,6] Eucalyptus oil if ingested is toxic.

Risk Assessment

Animals are unlikely to eat the leaves of the plant because of the strong pungent odor of the leaves. Most animal poisoning from *Melaleuca* arises from the application of the oil to the skin and hair coat as a means of cleaning the hair or as a treatment for various dermatologic diseases including ectoparasites.

Clinical Signs

Topical application of melaleuca oil or products containing the oil can cause ataxia, incoordination, muscle weakness, hypothermia, depression and behavioral abnormalities.[1,7] Severely affected animals may have elevations in liver enzymes, and blood urea nitrogen. Recovery can be expected if the animals can be bathed using mild soap to remove residual oil on the hair and skin. Activated charcoal should be administered orally to prevent further absorption of melaleuca oil that may have been ingested after animals have licked there haircoat. Melaleuca oil or other allergens in the plants are capable of inducing atopic allergic dermatitis in animals.[8]

References

1. Villar D, Knight MJ, Hansen SR, Buck WB: Toxicity of melaleuca oil and related essential oils applied topically on dogs and cats. Vet Hum Toxicol 36: 139-142, 1994.
2. Morris MC, Donoghue A, Markowitz JA, Osterhoudt KC: Ingestion of tea tree oil (Melaleuca oil) by a 4-year-old boy. Pediatric Emerg Care 19: 169-171, 2003.

3. Del Beccaro MA: Melaleuca oil poisoning in a 17-month-old. Vet Hum Toxicol 37: 557-558, 1995.
4. Jacobs MR, Hornfeldt CS: Melaleuca oil poisoning. Journal of Toxicology - Clinical Toxicology 32: 461-464, 1994.
5. Everist SL: Poisonous plants of Australia, 2nd ed. Angus & Robertson, London 546-548, 1981.
6. Webber JJ, Roycroft CR, Callinan JD: Cyanide poisoning of goats from sugar gums *(Eucalyptus cladocalyx)*. Austr Vet J 1985, 62: 28.
7. Bischoff K, Guale F: Australian tea tree *(Melaleuca alternifolia)* oil poisoning in three purebred cats. J Vet Diagn Invest 1998, 10: 208-210.
8. Mueller RS, Bettenay SV, Tideman L: Aero-allergens in canine atopic dermatitis in southeastern Australia based on 1000 intradermal skin tests. Austral Vet J 2000, 78: 392-399.

Melia azedarach

Family: Meliaceae

Figure 265 *Melia azedarach*

Figure 266 *Melia azedarach* flowers

Common Names
Chinaberry, Persian lilac, pride of India, Ceylon mahogany, Texas umbrella tree, white cedar, paraiso, piocha, beed tree.

Plant Description
Approximately10 species of the genus *Melia* are native to Asia, and have become widely distributed in the tropical regions of the world. The most ubiquitous species is *Melia azedarach*. It is a deciduous branched tree growing to heights of 15m, with dark or reddish brown bark. Leaves are twice-pinnately compound with 50 or more leaflets that are ovate to elliptic in shape. Inflorescences are loose panicles produced in the leaf axils. Individual flowers are fragrant white to lavender in color with 5-6 sepals and 5-6 petals. Fruits are fleshy, globular, drupes turning yellow or black when dry, and containing 1-6 seeds (Figures 265, 266, and 267).

Toxic Principle and Mechanism of Action
Several tetranortriterpenes referred to as meliatoxins are present throughout the plant, but especially in the fruits.[1]

Poisoning is only associated with the consumption of the berries. The mechanism of action of these toxins is poorly understood, and the toxicity of the berries varies considerably depending on the geographic area. Humans, cattle, pigs, rabbits, and dogs have most frequently been poisoned by eating Chinaberries.[2-4]

Figure 267 *Melia azedarach* berries

Risk Assessment
Chinaberries are commonly planted and can become invasive as the seeds are spread by birds in the tropics and mild temperate zones. It's rapid growth and fragrant flowers often make the Chinaberry an attractive garden tree. In some years, fruit production can be heavy and the fallen fruits are easily accessible to animals. Dogs appear to be particularly susceptible to fatal poisoning from eating Chinaberries.[3]

Clinical Signs
Increased salivation, vomiting, anorexia, and diarrhea occur initially, and may be followed by neurologic signs and paralysis. If sufficient quantities of the berries have been consumed animals may exhibit muscle weakness, ataxia, and seizures. Sudden death may be the only presenting sign in acute poisoning. Hepatic necrosis and degeneration of skeletal muscles may be detectable on histologic examination.

If an animal is witnessed eating Chinaberries, inducing vomiting may be the most efficient way to remove the berries. Activated charcoal administered orally is also effective. Sedatives and supportive fluid electrolyte therapy should be used in the more severely intoxicated animal.

References
1. Oelrichs PB, Hill MW, Valleyly VJ, MacLeod JK, Molinski TF: Toxic tetranor-triterpenes of the fruit of *Melia azedarach*. Phytochemistry 22: 531-534, 1983.
2. del Mendez MC, Elias F, Aragao M, Gimeno EJ, Riet-Correa F: Intoxication of cattle by the fruits of *Melia azedarach*. Vet Human Toxicol 44: 145-148, 2002.
3. Hare WR, Schutzman H, Lee BR, Knight MW: Chinaberry poisoning in two dogs. J Am Vet Med Assoc 210: 1638-1640, 1997.
4. Kwatra MS, Singh B, Hothi DS, Dhingra PN: Poisoning by *Melia azedarach* in pigs. Vet Rec 95: 421, 1974.

Family: Fabaceae

Figure 268 *Mimosa pudica*

Figure 269 *Mimosa pudica* leaves and flowers

Common Names
Sensitive plant, mimosa, touch-me-not, shame plant *(Mimosa pudica)*
The name mimosa is also commonly used for some species in the *Acacia* and *Albizia* genera

Plant Description
As a member of a large genus of some 450 species of tropical areas of Asia and the Americas, *Mimosa pudica* is native to Brazil, and has become widely cultivated as an ornamental for its sensitive leaves that fold when touched. A small, branching shrub, growing to 1 m in height, and with spiny stems. Leaves are fern-like, bipinnate, and fold up when touched (Figures 268 and 269). Inflorescences are globose, terminal, with numerous white or pink flowers. Fruits are leguminous pods.

Toxic Principle and Mechanism of Action
Mimosine is the principle toxicant in both *Mimosa* and *Leucaena* species. Once the plant tissues are damaged through chewing, the mimosine is degraded by plant enzymes to its toxic form which is an analogue and inhibitor of pyridoxine, an essential enzyme for DNA and RNA synthesis. Interference with DNA synthesis through the action of mimosine can result in characteristic hair loss from the ears, mane, and tail of cattle and horses and fleece loss in sheep.[1,2] Weakness, loss of appetite, enlarged thyroid glands, and ridges in the hoof wall have also been associated with mimosine toxicity. Toxicity has been reported in pigs and rabbits fed *Leucaena leukocephala*.[3]

Leucaena leukocephala (lead tree, jumby bean, white popinac) (Figure 270) is primarily a tree of the tropics, and has become established in Hawaii, Florida, and other Gulf States. It can become an invasive tree, and in some parts of the world is used as a source of forage for livestock because of its high protein content similar to that of alfalfa. Livestock once adapted to grazing the plant are able to degrade mimosine in the rumen and are not subject to poisoning.[2]

Risk Assessment
Sensitive plant *(M. pudica)* is not of concern as a toxic house plant as household

pets are unlikely to eat the plant in sufficient quantity to cause problems. The plant however can become an invasive weed when grown in tropical gardens and can escape to infest pastures grazed by livestock.

Clinical Signs
Hair loss from the ears, mane, and tail of cattle and horses and fleece loss in sheep is characteristic of mimosine toxicity. Weakness, loss of appetite, enlarged thyroid glands, and ridges in the hoof wall have also been associated with mimosine toxicity.

Figure 270 *Leucaena leukocephala* (inset-seed pods)

Treatment is seldom necessary as the toxicity is self limiting once the animals are provided other forages.

References
1. Anderson RC et al: Drought associated poisoning of cattle in South Texas by the high quality forage legume *Leucaena leucocephala*. Vet Hum Toxicol 43: 95-96, 2001.
2. Hammond AC: Leucaena toxicosis and its control in ruminants. J Anim Sci 73: 1487-1492, 1995.
3. *Owen LN*: Hair loss and other toxic effects of *Leucaena glauca* ("jumbey"). Vet Rec 70: 454-456, 1958.

Mirabilis

Family: Nyctaginaceae

Common Names
Four o'clocks, Marvel of Peru, umbrellawort *(Mirabilis jalapa)*

Plant Description
Consisting of about 50-60 species native to Central and South America, *Mirabilis* are annuals or perennial herbaceous plants arising from fleshy or woody roots. Depending on the species, the branching plants may reach heights of 100 cm, and have dark green, opposite, ovate to

Figure 271 *Mirabilis jalapa*

cordate, petiolate to sessile leaves. Flowers are produced terminally, and are funnelform, in a variety of colors, and opening in the late afternoon (Figures 271 and 272). The fruits are 5 sided achenes.

Figure 272 *Mirabilis jalapa*

Toxic Principle and Mechanism of Action
The seeds and roots contain the alkaloid trigonelline which is an irritant to the skin and the digestive tracts.

Risk Assessment
The only reported cases of poisoning have involved the common four o'clock (*M. jalapa*).[1] This common garden plant is a prolific seed producer and can therefore be a potential source of the toxic seeds for children or pets.

Clinical Signs
Chewed seeds if swallowed can cause vomiting and diarrhea.

References
1. O'Leary SB: Poisoning in man from eating poisonous plants. Arch Environ Health 9: 216-242, 1964.

Momordica
Family: Cucurbitaceae

Figure 273 *Momordica charantia*

Common Names
Balsam pear, balsam apple, bitter gourd, bitter cucumber, balsamina
Two species are naturalized in North America; *M. balsamina* and *M. charantia*.

Plant Description
Comprising a genus of 35-40 species native to the tropical areas of Africa, *Momordica* species have become naturalized in tropical areas of North America.
The plants are vines, with slender, hairy, grooved, stems that may grow to 10m in length. The leaves are simple, triangular, ovate or cordate, with 3-7 palmate lobes. Flowers are single, showy, yellow to orange in color, with 5 sepals and 5 petals. Fruits are up to 6cm in length, pendulous, and covered with pointed, wart-like projections, turning yellow when ripe and splitting open to reveal numerous seeds with bright orange-red fleshy arils (Figure 273).

Toxic Principle and Mechanism of Action
Tetracyclic cucurbitane terpenoids (momordicin I,II), and at least 2 lectins are the toxins most responsible for the toxicity of the plant.[1,2] All parts of the plant are toxic with the exception of the fleshy ripe arils covering the seeds.[3] The

compounds act as gastrointestinal irritants and in high doses may cause depression, muscle tremors, and seizures. Similar cucurbitane compounds are found in bryony (*Bryonia* spp.) and buffalo gourd *(Cucurbita foetidissima)*, species of which are native to North America. Mormordica species have found wide use in herbal remedies and may have benefit in treating certain cancers.[4]

Risk Assessment
Since fruits and seeds are attractive to some birds, the plant seeds are quite easily spread. The brightly colored fruits and seeds are also attractive to children and to pets especially where the vines are grown in gardens and along fences. *Momordica* poisoning is not frequently reported in dogs except in Florida where several instances of poisoning have occurred.[3]

Clinical Signs
Vomiting, salivation, abdominal pain, and diarrhea are the most likely signs associated with *Mormordica*. Dogs that eat the ripe fruits reportedly develop convulsions and vomiting.[3]

Severely affected animals may require sedation and intravenous fluid therapy to counteract dehydration and shock following severe diarrhea and vomiting.

References
1. Connolly JD, Hill RA: Dictionary of Terpenoids. Vol 2. Di- and higher terpenoids. Chapman & Hall, London 1173-1180, 1991.
2. Lin J-Y, Hou M-J, ChenY-C: Isolation of toxic and non-toxic lectins from the bitter-pear melon *Momordica charantia*. Linn. Toxicon 653-660, 1978.
3. Morton JF. Plants Poisonous to People in Florida. 2nd ed. JF Morton ed. Southeastern Printing, Stuart, Florida. 73-74, 1982.
4. Grover JK, Yadav SP: Pharmacological actions and potential uses of *Momordica charantia*: a review. J Ethnopharmacology. 93: 123-132, 2004.

Monstera
Family: Araceae

Common Names
Ceriman, fruit salad plant, bread fruit, Swiss cheese plant, window leaf plant, hurricane plant.

Plant Description
Consisting of some 25 species originating in tropical America, the genus is mostly commonly represented by *Monstera deliciosa*. These large evergreen perennial climbers, usually with long aerial roots become epiphytic as they climb trees. The juvenile leaves are usually small and entire,

Figure 274 *Monstera deliciosa*
(Inset are the fruits) Photographer: Jenger Smith

while the mature leaves are much larger and become perforated (Figure 274). The creamy-white flower spikes are typically enclosed in a large spathe to 18 in. (45cm) in length. The ripe fruits of some species are edible often taking as long as a year to ripen.

Monstera deliciosa is a common tropical garden plant, and is frequently grown as a potted house plant. This species is often mistaken for the similar split-leaf philodendron.

Toxic Principle and Mechanism of Action
Like other members of the Araceae family, *Monstera* species contain oxalate crystals in the stems, leaves, and aerial roots that have the potential of causing contact irritation especially if the plant is chewed.[1,2] (see *Dieffenbachia* spp.) The ripe fruits are edible, but the unripe fruit causes marked irritation to the throut of people eating it.

Risk Assessment
Monstera species pose little risk to animals, even though the plants are commonly grown in tropical gardens and as house plants.

Clinical Signs
Salvation and vomiting are potential problems if animals eat the plants. Treatment is seldom necessary. The plants should be removed or made inaccessible to the animals eating them.

References
1. Genua JM, Hillson CJ: The occurrence, type, and location of calcium oxalate crystals in the leaves of 14 species of Araceae. Ann Bot 56: 351-361, 1985.
2. Franceschi VR, Horner HT: Calcium oxalate crystals in plants. Bot Rev 46: 361-427, 1980.

Nandina domestica

Family: Berberidaceaea

Common Names
Nandina, Chinese sacred bamboo, heavenly bamboo

Plant Description
Primarily a plant of China and Japan, the single species, *Nandina domestica*, has many cultivars. It is favored for its delicate flowers and bright red-orange berries. It is not a true bamboo but rather is closely related to *Podophyllum* species (May apple), and *Berberis* species (Barberry). As evergreen or deciduous woody shrubs, Nandina have erect stems, 2-3 times pinnately compound leaves with elliptic leaflets. Leaves turn red-purple in the Fall. The inflorescence is a panicle with numerous, small, cream-white flowers. Berries are orange-red or white with 1-3 seeds (Figures 275-277).

Figure 275 *Nandina domestica* flowers

Toxic Principle and Mechanism of Action
Some cultivars of Nandina contain significant quantities of cyanogenic glycosides (hydroxymandelonitrile) which when hydrolyzed release hydrogen cyanide.[1] Also present are various protoberberine alkaloids of unknown toxic significance. The best known of the

Figure 276 *Nandina domestica*

alkaloids is berberine that is known to have anticholinesterase activity and causes smooth muscle relaxation and hypotension. In higher doses seizures may occur, possibly as a result of the antagonistic effect of nantenine on serotonin.[2]

Risk Assessment
Commonly grown as an ornamental garden plant in many areas of North America, the plant is rarely a problem to household pets. However, the bunches of red berries (or white) that persist on the bushes in winter are attractive to animals. A brief report is given of a puppy that developed seizures after eating the berries.[1] Ruminants are more likely to be at risk from eating Nandina because they more readily hydrolyse the cyanogenic glycosides to hydrogen cyanide than do simple stomached animals. Prunings from Nandina should not be fed to ruminants, nor should it be planted in or around livestock enclosures.

Figure 277 *Nandina domestica alba*

Clinical Signs

Seizures appear to be the dominant sign of poisoning in dogs.[2] However, in the Nandina cultivars with high cyanide content, the clinical signs will relate to those associated with acute anoxia caused by the hydrogen cyanide. Acute onset dyspnea, cherry-red colored mucous membranes and venous blood, and death within a few hours of eating a toxic dose of the plant can be anticipated, especially in ruminants.

If cyanide poisoning is suspected, the animal should be treated intravenously with sodium thiosulfate solution. (see *Prunus* spp.)

References

1. Bradley M, Nieman LJ, Burrows GE: Seizures in a puppy. Vet Hum Toxicol 30: 121, 1981.
2. Shoji N, Umeyama A, Takemoto T, Ohizumi Y: Serotonergic receptor antagonist from *Nandina domestica* Thunberg. J Pharm Sci 90: 723, 1984

Narcissus

Family: (Amaryllidaceae)

Figure 278 *Narcissus* spp. (Paperwhite)

Common Names

Daffodil, narcissus, jonquil, paperwhite

Plant Description

The genus *Narcissus* consists of approximately 60 species that originate from Europe, North Africa, and western Asia. A vast number of cultivars have been developed and are grouped horticulturally into 12 divisions based upon the size and shape of the trumpet and the petals. Plants develop from bulbs that vary considerably in size and shape, but are generally ovoid with brown papery membranes. The leaves are basal direct or spreading, with narrow blades. Flowers are either single or in clusters of up to 20 flowers and flowers are generally large and showy, fragrant or not fragrant, and ranging in color from white to yellow or orange (Figures 278-281).

Toxic Principle and Mechanism of Action

At least 15 phenanthridine alkaloids including lycorine, have been identified in the leaves, stems, and bulbs of *Narcissus*.[1] The concentrations of the alkaloids are highest in the outer layers of the bulbs. The total alkaloid concentration in the leaves and parts of the bulbs is reported as 0.5%. Phenanthridine alkaloids have been isolated from other genera of the Amaryllis family, including species of *Amaryllis, Clivia, Cooperia, Eucharis, Galanthus, Hippeastrum, Haemanthus, Hymenocallis, Leucojum, Nerine, Sprekelia, and Zephranthes.*

Figure 279 *Narcissus* - Daffodil

The bulbs also contain calcium oxalate raphides that are largely responsible for the contact irritant dermatitis, that is common in people who handle the bulbs frequently. The calcium oxalate crystals penetrate the skin, allowing allergenic substances such as the alkaloids masonin and homolycorin to enter and stimulate an allergic response.[2,3] The alkaloids have emetic, hypotensive, and respiratory depressant effects, and cause excessive salivation, abdominal pain, diarrhea, and hypotension. Calcium oxalate raphides may also contribute to the digestive symptoms especially if the bulbs or plants are chewed. There are other compounds in Narcissua species that have been shown to have antiviral, antimitotic, and antitumor properties.

Figure 280 *Narcissus* 'Kilworth'

Risk Assessment

Being one of the most common garden and potted houseplants, Narcissus species are readily accessible to household pets. Usually purchased as a leafless bulb, and if not planted immediately, the bulb can be something puppies like to chew on. Cats may eat the dried stems and leaves and become intoxicated.[4] Most reported cases of daffodil (*Narcissus* spp.) poisoning occur in people who eat the bulbs.[3,5]

Figure 281 *Narcissus* bulbs

Clinical Signs

Vomiting, excessive salivation, abdominal pain, diarrhea, and difficulty in breathing, are associated with the phenanthridine alkaloids present in the lily family. If large quantities of the leaves, stems, and bulb are consumed, depression, ataxia, seizures, bradycardia, and hypotension may develop. Poisoning is rarely fatal, and can generally be treated symptomatically.

References
1. Martin SF: The Amaryllidaceae alkaloids. In The Alkaloids: Chemistry and Physiology. vol 30, Brossi A (ed) Academic Press, San Diego, Calif 1987, pp 251-376, 1987.
2. Gude M, Hausen BM, Heitsch H, Konig WA: An investigation of the irritant and allergenic properties of daffodils (*Narcissus pseudonarcissus* L., Amaryllidaceae). A review of daffodil dermatitis. Contact Dermatitis 19: 1-28, 1988.
3. Spoerke DG, Smolinske SC: (Eds). In Toxicity of Houseplants. CRC Press, Boca Raton pp 172-174, 1990.
4. Saxon-Buri S: Daffodil toxicosis in cat. Can Vet J 45: 248-250, 2004.
5. Vigneau CH, Tsao J, Chamaillard C, Galzot J: Accidental absorption of daffodils *(Narcissus jonquilla)*: two common intoxications. Vet Hum Toxicol 1982, 24: 35, 1982.

Nepeta cataria
Family: Lamiaceae

Figure 282 *Nepeta cataria*

Common Names
Catnip, catmint, catwort, field balm

Plant Description
As a member of a large genus of some 220 species, *Nepeta cataria* originates in Eurasia, and has now become established as a weedy perennial in many areas of eastern and Central North America. Growing to 1m in height, it is an erect, branching plant, with square stems arising from a strong taproot. Leaves are opposite, simple, cordate to narrowly triangular, with coarsely serrate margins, and the underside of the leaves being more hairy than the top side (Figure 282). Inflorescence is a terminal spike with individual flowers white to pink with red to purple markings. Fruits are ovoid nutlets containing one seed each.

Toxic Principle and Mechanism of Action
The monoterpene nepetalacetone is the major component of catnip responsible for the attraction of cats.[1] After inhalation and/or ingestion of the plant or catnip oil, cats may exhibit a range of pleasurable and hallucinogenic-like signs for up to 10 minutes, after which the cat loses interest in the plant and the animal returns

to normal behavior.[2,3] These pleasurable and hallucinatory responses are similar to those encountered in people smoking catnip, marijuana, and lysergic acid diethylamide (LSD.)[4] Similarities exist in the chemical structure of nepetalacetone and tetrahydrocannabinol in marijuana, and LSD.

Risk Assessment
There is minimal if any risk to cats eating catnip as interest in the plant is of short duration.

Clinical Signs
A variety of signs can be seen in cats when they first encounter the plant or toys that contain the dried plant. Intense interest, sniffing, and chewing the plant is followed by salivation, rubbing or rolling on the plant, clawing, shaking the catnip toy, vigorous jumping and playing with the toy, followed by intense grooming and licking of the genitalia.[3] Some cats show apparent hallucinations as if there were phantom mice or butterflies to chase. Male cats may develop penile erections and female cats may display estrous behavior briefly.[3] Typically cats will suddenly lose interest in the plant or catnip toy after about 10 minutes.

References
1. Regnier FE, Eisenbraun EJ, Waller GR: Nepetalacetone and epinepetalacetone from *Nepeta cataria.* Phytochem 6: 1271-1280, 1967.
2. Harney JW. Barofsky IM. Leary JD. Behavioral and toxicological studies of cyclopentanoid monoterpenes from *Nepeta cataria.* Lloydia 41: 367-374, 1978.
3. Hatch RC. Effect of drugs on catnip *(Nepeta cataria)*-induced pleasure behavior in cats. Am J Vet Res 33: 143-155, 1972.
4. Jackson B, Reed A: Catnip and the alteration of consciousness. J Am Med Assoc 207: 1349-1350, 1969.

Nerine
Family: Liliaceae (Amaryllidaceae)

Common Names
Guernsey lily, nerine lily, spider lily.

Plant Description
Consisting of about 30 species indigenous to Southern Africa, *Nerine* species are lilies growing from bulbs. White to rosy pink, narrow-petalled, trumpet-shaped, showy flowers are produced in the fall on tall stems before the basal grass-like leaves emerge (Figures 283 and 284).

Figure 283 *Nerine bowdenii*

Figure 284 *Nerine bowdenii* flowers

Toxic Principle and Mechanism of Action

Several phenanthridine alkaloids including lycorine, neronine, and tazettine have been identified in the leaves, stems, and bulbs of *Nerine*.[1] The phenanthridine alkaloids are present in many of the Liliaceae, most notably in the Narcissus group (see *Narcissus*). The alkaloids have emetic, hypotensive, and respiratory depressant effects, and cause excessive salivation, abdominal pain, and diarrhea. As many as 15 other phenanthridine alkaloids have been isolated from other genera of the Amaryllis family, including species of *Clivia, Galanthus, Haemanthus, Hippeastrum, Hymenocallis, Leucojum, Narcissus, Sprekelia,* and *Zephranthes*.[1]

Risk Assessment

This attractive garden plant, and occasionally potted houseplant, has not been reported to cause poisoning in animals, but it has the potential to do so.

Clinical Signs

Vomiting, excessive salivation, abdominal pain, diarrhea, and difficulty in breathing are associated with the phenanthridine alkaloids present in the lily family. If large quantities of the leaves and bulb are consumed, depression, ataxia, seizures, and hypotension may develop. Poisoning is rarely fatal, and can generally be treated symptomatically.

References

1. Martin SF: The Amaryllidaceae alkaloids. In The Alkaloids: Chemistry and Physiology. vol 30, Brossi A (ed) Academic Press, San Diego, Calif 251-376, 1987.

Nerium oleander

Family: Apocynaceae

Common Names

Oleander, rose laurel, laurel Colorado

Plant Description

Consisting of a single species with multiple cultivars, *Nerium oleander* is a native of the Mediterranean area and tropical Asia, and is widely cultivated in the warmer regions of the world. It is a popular landscaping plant because it tolerates relatively dry conditions. Oleander is commonly used in hedges and in highway landscaping.

A perennial evergreen branching shrub that can attain heights of 15-20 ft (6 metres), with simple, dark green, glossy, leathery, lanceolate, whorled leaves, with a prominent mid-rib. The fragrant showy flowers are produced terminally on branches, and are funnel shaped with 5 petals, in colors of white, red, or pink (Figures 285 and 286). Some cultivars have double petals. Fruits are bean-like seed pods with numerous plumed seeds.

Figure 285 *Nerium oleander* (flowers and pod)

Figure 286 *Nerium oleander*

Toxic Principle and Mechanism of Action

Nerium oleander contains numerous cardenolides and their genins that are concentrated in the leaves, flowers, and seeds.[1,2] Also present in the plant are terpenoids that possibily account for the gastrointestinal irritation seen with oleander poisoning.

The cardiotoxic effect of the oleander cardenolides is similar to that caused by digitoxin and digoxin found in the *Digitalis* species. The primary action of the cardenolides is on the cell membrane, where interference with normal transport of sodium and potassium ions across the cell membrane occurs allowing an influx of calcium.[3] At low doses, myocardial function may improve, but at high doses cardiac conduction is impaired with resulting arrhythmias, heart block, and death.

A wide variety of animals including humans, dogs, cats, horses, cattle, sheep, goats, llamas, and birds have been poisoned by oleander.[4-10]

Risk Assessment

Oleander is a common plant in many gardens and is frequently used in landscaping in tropical and subtropical areas. In temperate climates it is often sold as a potted plant for indoor use. Considering that oleander is one of the most cardiotoxic plants known, and is poisonous to most animals including humans, it should not be planted where it could be a risk to children or household pets. It should not be planted in or around animal enclosures, and the leaves and branches pruned from oleander shrubs should never be fed to animals.[11] Oleander is highly poisonous to birds and therefore should not be included in aviaries.[8,11] Compost made from oleander leaves can result in detectable but low levels of the glycoside oleandrin in plants mulched with the oleander compost.[12]

Clinical Signs

Excessive salivation, vomiting, and diarrhea are commonly seen initially in dogs, cats and most other species poisoned by oleander. The diarrhea may contain blood. Within a few hours of ingesting the plant, cardiac signs develop including weakness, depression, irregular pulse, bradycardia, and increased respiratory rate. Electrocardiographically, S-T depression, bradycardia, extrasystoles, and various dysrhythmias will be apparent. Hyperkalemia may or may not be present. Depending on the quantity of the cardenolides ingested, animals may exhibit signs of depression and heart irregularity for many hours before recovering or they may die suddenly due to cardiac arrest.

At postmortem examination, there are generally no specific lesions present. Animals that survived for several days often have necrosis of the myocardium. A diagnosis of oleander poisoning can be made by finding the distinctive leaf parts in the animal's stomach contents, and by detection of the cardenolides in the stomach contents using high pressure liquid chromatography (HPLC) methods.[13] Successful treatment of oleander poisoning depends on early recognition of the toxicity.

Induction of vomiting, gastric lavage, and/or the oral administration of activated charcoal is appropriate for removing the plant and preventing further absorption of the toxins. Cathartics may also be used to help eliminate the plant rapidly from the digestive system. Serum potassium levels should be closely monitored and appropriate intravenous fluid therapy initiated as necessary. Phenytoin, as an anti-arrhythmic drug effective against supraventricular and ventricular arrhythmias, can be used as necessary. Similarly, atropine and propanalol have been used. The use of commercially available digitalis-specific antibody (Digibind – Burroughs Wellcome) may be a beneficial in counteracting the effects of the cardenolides.[14-16]

References
1. Paper D, Franz G: Glycosilation of cardenolides aglycones in leaves of Nerium Oleander. Planta Medica 55: 30-34, 1989.
2. Radford DJ, Gillies AD, Hinds JA, Duffy P: Naturally occuring cardiac glycosides. Medical J Australia. 144: 540-544, 1986.
3. Ooi H, Colucci: Digitalis and allied cardiac glycosides. In Goodman and Gilman's The Pharmacological Basis of Therapeutics, 10th ed, hardman JG,

Linbird LEP eds, McGraw-Hill, New York pp 916-921, 2001.

4. Galey FD, Holstege DM, Johnson BJ, Siemans L: Toxicity and diagnosis of oleander *(Nerium oleander)* poisoning in livestock. In Toxic Plants and Other Natural Toxicnts, Garland T, Barr AC eds. New York pp 215-219, 1998.

5. Ada SE, Al-Yahya MA, Al-Farhan AH: Acute toxicity of various oral doses of dried Nerium oleander leaves in sheep. Am J Chinese Med 29: 525-532, 2001.

6. Langford SD, Boor PJ: Oleander toxicity: an examination of human and animal toxic exposures. Toxicology 109: 1-13, 1996.

7. Alfonso HA, Sanchez LM, Merino N, Gomez BC: Intoxication due to *Nerium oleander* in geese. Vet Human Toxicol 1994, 36: 47.

8. Shropshire CM, Stauber E, Arai M: Evaluation of selected plants for acute toxicosis in budgerigars. J Am Vet Medical Assoc. 200: 936-9, 1992.

9. Clark RF. Selden BS. Curry SC. Digoxin-specific Fab fragments in the treatment of oleander toxicity in a canine model. Ann Emerg Med. 20: 1073-1077.

10. Shaw D, Pearn J: Oleander poisoning. Med J Aust 1979, 2: 267-269.

11. Arai M, Stauber E, Shropshire CM: Evaluation of selected plants for their toxic effects in canaries. J Am Vet Med Assoc 1992, 200: 1329-1331.

12. Downer J, Craigmill A, Holstege D: Toxic potential of oleander derived compost and vegetables grown with oleander soil amendments. Vet Hum Tox 2003, 45: 219-221.

13. Tor ER, Holstege DM, Galey FD: Determination of oleander glycosides in biological matrices by high-performance liquid chromatography. J Agric Food Chem 1996, 44: 2716-2719.

14. Shumaik GM, Wu AW, Ping AC: Oleander poisoning: treatment with digoxin-specific Fab antibody fragments. Ann Emerg Med 1988, 17: 732-735.

15. Gfeller RW, Messonier SP: Handbook of small animal toxicology and poisoning 2nd ed. Mosby, St. Louis, Missouri. 2004, pp 161-162.

16. Smith TW et al: Treatment of life-threatening digitalis intoxication with digitoxin-specific Fab antibody fragments. Experience and 26 cases. N Eng J Med 1982, 307: 1357-1362.

Family: Solanaceae

Figure 287 *Nicotiana sylvestris*

Figure 288 *Nicotiana alata* hybrid

Common Names
Tobacco, burley tobacco – *N. tabacum*
Tree tobacco, mustard tree – *N. glauca*
Flowering tobacco – *N. alata*

Plant Description
A genus of 67 species of annuals, biennials, and perennials, *Nicotiana* are native to America and Australia. Small erect herbs to small trees, with large, basal, lanceolate to ovate, often aromatic, sessile or petiolate, sticky leaves. Inflorescences are terminal panicles or racemes. Flowers are showy, usually fragrant, and opening in the evening, although hybrid varieties bloom during the day. Flowers are 5 lobed, tubular or funnel-shaped, and in a variety of colors including white, pink, red, and yellow (Figures 287, 288, and 289). Fruits are globular capsules containing many seeds.

Toxic Principle and Mechanism of Action
A variety of pyridine and piperidine alkaloids are present in *Nicotiana* species, the most toxic of which are nicotine and anabasine. Nicotine is much more toxic than anabasine and is a rapidly acting depolarizing agent of sympathetic and parasympathetic ganglia.[1] It also directly affects the brain. In low doses nicotine causes stimulation, while at high doses it causes paralysis. The minimum lethal dose of nicotine in dogs and cats is 20-100mg.[2] Anabasine is of primary interest as it is teratogenic causing fetal deformities such as cleft palate.[3,4]

Risk Assessment
As garden plants grown for their showy and fragrant flowers, cultivated *Nicotiana* species are rarely a source of poisoning to animals. Poisoning however may occur if large amounts of garden trimmings are dumped where livestock might have access to them, or when livestock are fed *N. tabacum* stalks after the leaves have been harvested for tobacco production.

Most cases of nicotine poisoning in dogs occur when they eat cigarettes, cigars, chewing tobacco (sweetened), nicotine gum, nicotine patches and other products containing nicotine such as insecticide dusts or sprays.

Clinical Signs
Bradycardia and slow respiratory rates are seen initially, but as the toxic dose increases, excessive salivation, urination, defecation, increased heart and respiratory rates, dilated pupils, muscle tremors, incoordination, weakness, and

Figure 289 Nicotiana alata

collapse may be seen. Death from a lethal dose usually occurs as a result of respiratory and cardiac failure.

Treatment for nicotine poisoning is symptomatic. Activated charcoal orally helps to reduce further absorption of nicotine from the digestive tract. Antacids should not be given as they increase the absorption of nicotine.[5] Intravenous fluid therapy, and urine acidification help to hasten the excretion of nicotine and its metabolites through the urine. Patients should be closely monitored and treated with oxygen, positive pressure respiration, and sedatives to control seizures as necessary.[5]

References
1. Adams HR: Cholinergic pharmacology: Autonomic drugs. In Veterinary Pharmacology and Therapeutics, 7th ed. Adams HR ed. Iowa State University Press, Ames, Iowa. 131-140, 1995.
2. Vig M: Nicotine poisoning in a dog. Vet Hum Toxicol 32: 573-57, 1990.
3. Keeler RF et al: Teratogenic effects of Nicotiana glauca and concentration of anabaosine, the suspected teratogen in plant parts. Cornell Vet 71: 47-53, 1981.
4. Panter KE, Keeler RF: Induction of cleft palate in goats by Nicotiana glauca during a narrow gestational period, and the relation to reduced fetal movement. J Nat Toxins 1: 25-32, 1992.
5. Plumlee KH: Nicotine. In Small Animal Toxicology Peterson ME, Talcott PA eds. WB Saunders, New York. 600-602, 2001.

Family: Liliaceae (Hyacinthaceae)

Figure 290 *Ornithogallum saundersiae*

Common Names
Star of Bethlehem, chinkerinchee,

Plant Description
A large genus of some 150 species native to Africa, Europe, and Asia, *Ornithogallum* species are stemless lilies arising from a bulb. The basal leaves are linear and fleshy. The inflorescence consists of short or tall, pyramid-shaped racemes of white, cupped or star-shaped flowers, and fragrant depending on the species (Figures 290 and 291). Fruits are capsules with many seeds. (Some species such as *O. umbellatum* (Star of Bethlehem) are invasive) (Figure 292).

Toxic Principle and Mechanism of Action
All parts of the plant contain a variety of toxic cardenolides similar to those found in lily of the valley *(Convallaria majalis)*.[1,2] A variety of saponins are also present in the plant and may account for some of the gastrointestinal signs. The bulb is particularly toxic to many animal species including humans, dogs, and livestock.[2,3]

Figure 291 *Ornithogallum saundersiae* flowers

Risk Assessment
Ornithogallum species are not common in North America, although species such as *O. umbellatum* has been introduced into the south eastern States where it has become invasive in pastures and meadows in some areas. The attractive species *O. thyrsoides* is sometimes grown as a garden or potted plant for its attractive, showy flowers.

Clinical Signs
Sudden death or gastrointestinal signs characterize poisoning from *Ornthogallum* species.[2] Blindness has been reported in some animals after eating the plant.

References

1. Ghannamy U, Kopp B, Robien W, Kubelka W: Cardenolides from *Ornithogallum boucheanum*. Planta Med 53: 172-178, 1987.
2. Kellerman TS, Coetzer JAW, Naude TW: Plant Poisonings and Mycotoxicoses of Livestock in Southern Africa. Oxford University Press, Capetown pp 141-144, 1990.
3. Bamhare C: Suspected cardiac glycoside intoxication in sheep and goats in Namibia due to *Ornithogalum nanodes* (Leighton).Onderstepoort J Vet Res 65: 25-30, 1988.

Figure 292 *Ornithogallum umbellatum* (Star of Bethlehem)

Oxalis

Family: Oxalidaceae

Common Names
Shamrock, wood sorrel, soursob, Bermuda buttercup, oxalis
Irish shamrock – *O. acetosella*

Plant Description
Comprising over 800 species, *Oxalis* are found widely in many parts of the world. Most are wild flowers or weeds, but some are cultivated as garden and potted indoor plants for their attractive foliage and flowers. Annual or perennial herbs growing from scaly bulbs, rhizomes, or taproots. Stems are thin, decumbent or erect, up to 40cm in height, and haired or hairless. Leaves are palmate, cordate, with 3 leaflets that tend to fold at night. The sap and plant parts are sour to the taste. Infloresences are single flowers or cymes or umbels. Flowers are bell-shaped, 5 petalled, and in various colors ranging from white to yellow and pink to purple (Figures 293-295). The flowers fold up at night. Fruits are capsules that are explosively dehiscent , scattering numerous seeds.

Figure 293 *Oxalis* hybrid

Figure 294 *Oxalis* hybrid

Figure 295 *Oxalis purpurea*

Toxic Principle and Mechanism of Action
Soluble oxalates are present in all parts of the plant and are responsible for the sour taste of the plants. The sap is acidic due to the presence of potassium oxalate and free oxalic acid, with total oxalate in some species being 16% dry matter.[1] If eaten in quantity the oxalates can induce hypocalcemia and oxalate nephrosis. Most poisoning has occurred in sheep that graze stands of *Oxalis pes-caprae* in Australia.[2] Goats have also been affected.[3]

Risk Assessment
It is unlikely that household pets or children would eat much of the plant because of its sour taste.

Clinical Signs
Sheep that have eaten large amounts of the plant become depressed, ataxic and recumbent because of hypocalcemia following the formation of insoluble calcium oxalate. Oxalate crystals may be seen in the urine and animals die from an oxalate nephrosis.

In the event oxalate poisoning is confirmed, the intravenous administration of calcium borogluconate may rapidly reverse hypocalcemic effects, but oxalate nephrosis may not be affected and animals may die several days later from renal failure.

References
1. Libert B, Franceschi VR: Oxalate in crop plants. J Agric Food Chem 35: 926-938, 1987.
2. McIntosh GH: Chronic oxalate poisoning in sheep. Austr Vet J 48: 535, 1972.
3. Rekhis J, Amara A: Two cases of food poisoning by *Oxalis cernua* in goats. Rev Vet Med 141: 8-9, 1990.

Family: Papaveraceae

Common Names

Icelandic or artic poppy	*Papaver nudicaule*
Oriental poppy	*P. orientale*
Field or Flanders poppy	*P. rhoeas*
Opium or carnation poppy	*P. somniferum*
Sea or horned poppy	*Glaucium* spp.

Figure 296 *Papaver somniferum*

Plant Description

There are 70 species of annual, biennial, or perennial poppies that are native to temperate climates especially in Europe, Africa, Asia, North America, and Australia. Numerous cultivars have been developed. Arising from a taproot, the stems are erect, branched or unbranched, thornless, often with hairs, and ranging up to 150cm in height. Leaves are basal, alternate, hairy, variably pinnately lobed, and with entire or toothed edges. Flower buds are pendent. Flowers are showy being carried on long wiry stems. Sepals 2-3, petals 4-6, in colors of white, yellow, orange, red, and purple. Flowers are usually short lived and are followed by the distinctive seed capsules that contain numerous pitted seeds. The viscous sap ranges in color from white to orange-red (Figures 296-300).

Figure 297 *Papaver rhoeus*

Toxic Principle and Mechanism of Action

The phenanthrene alkaloids including morphine, heroin, papaverine, and codeine, are principally found in the opium poppy, opium being the crude extract of a mixture of alkaloids obtained from the sap. Numerous isoquinoline

Figure 298 *Papaver croceum*

alkaloids are present in other species of Papaveraceae.[1,2] All parts of the plants and especially in the sap and seeds contain the alkaloids. The seeds from opium poppies if eaten in quantity can result in detectable opiate levels in the urine.[3]

Figure 299 *Papaver somniferum* double flower

Figure 300 *Papaver somniferum* capsule

The quantity and type of alkaloids present varies with the species of poppy, but all contain sufficient amounts to be considered toxic.

The *Papaver* alkaloids have a significant effect on the nervous system being agonists of the opioid receptor thereby causing a variety of effects ranging from drowsiness, depression, pain relief, euphoria, excitement, and coma depending on the dose consumed and the species of animal. High doses will cause miosis and severe respiratory depression. The alkaloids also affect the digestive system causing vomiting, decreased intestinal motility, and constipation.

Risk Assessment
Poisoning from poppies is rare in animals except where the plant trimmings are accidentally fed to livestock and horses. Poppies are more palatable when wilted.

Clinical Signs
In animals the signs of *Papaver* poisoning are usually mild and include excitement, ataxia, loss of appetite, decreased respiration, staring expression, drowsiness, and deep sleep.[4-7]
Treatment is seldom necessary as signs are self limiting.

References
1. Preininger V. Chemotaxonomy of Papaveraceae and Fumariaceae. In The Alkaloids vol 29, Brossi A ed. Academic Press, Orlando Florida. 1986, 1-89.
2. Yasmin R, Naru AM. Biochemical analysis of *Papaver somniferum* (opium poppy). Biochemical Society Transactions. 19: 436S, 1991.
3. Lo DS. Chua TH. Poppy seeds: implications of consumption. Medicine, Science & the Law. 32: 296-302, 1992.
4. Malmanche I. Suspected *Papaver nudicaule* (Icelandic poppy) poisoning in two ponies. NZ Vet J. 18: 96-97, 1970.
5. McLennan GC. Poisoning of sheep by ingestion of Icelandic poppies *(Papaver nudicaule)*. Aust Vet J. 5: 117, 1929.
6. Odendaal JS. Suspected opium poisoning in two young dogs. J South African Vet Assoc. 57: 113-114, 1986.
7. Burrows GE, Tyrl RJ. Toxic Plants of North America. Iowa State University Press, Ames. 850-856, 2001.

Parthenocissus

Family: Vitaceae

Common Names
Virginia creeper *(Parthenocissus quinquefolia)*
Boston ivy *(P. tricuspidata)*

Plant Description
Consisting of some 10-15 species that are native to North America and Asia, *Parthenocissus* are perennial, deciduous, woody vines, with tendrils with terminal pads that enable the creeper to climb on walls, trees, and fences. The leaves are palmate, compound with 5 leaflets that are serrate. The leaves turn bright red in the Fall. Small greenish flowers are produced in paniculate cymes. Fruits are blue to black berries with 1-4 seeds (Figures 301 and 302).

Toxic Principle and Mechanism of Action
The toxin present in *Parthenocissus* is not known. Oxalate raphides, and possibly other compounds, may be responsible for the signs of gastroenteritis reported in children eating the leaves or berries.[1,2] Similar toxicity has been reported in budgerigars fed the leaves.[3]

Figure 301 *Parthenocissus quinquefolia*

Figure 302 *Parthenocissus quinquefolia* fruits

Risk Assessment
Virginia creeper is commonly grown for its ability to rapidly cover walls and fences. Once established, the vines often produce heavy crops of attractive blue-black berries in the Fall that drop to the ground and are a potential risk to pets and children.

Clinical Signs
Children eating the leaves or berries have developed mild colic, vomiting, and diarrhea.

References
1. O'Leary SB, Poisoning in man from eating poisonous plants. Arch Environ Health 9: 216-242, 1964.
2. Lampe KF, McCann MA. AMA Handbook of Poisonous and Injurious Plants. Am Med Assoc, Chicago, Illinois pp 197, 1985.
3. Shropshire CM, Stauber E, Arai M: Evaluation of selected plants for acute toxicosis in budgerigars. J Am Vet Med Assoc 200: 936-939, 1992.

Pedilanthus

Family: Euphorbiaceae

Figure 303 *Pedilanthus tithymaloides*

Common Names

Slipper flower, Devil's backbone, Jacob's ladder, Christmas candle, redbird cactus, candelillo,

Plant Description

A genus of 14 species of succulent, perennial, evergreen or deciduous shrubs native to the tropical and desert areas of North and South America, and the Caribbean area. There are 2 species in North America; *P. tithymaloides* and *P. macrocarpus*. Stems are erect, up to 3m in height, jointed, and contain a milky sap. Leaves are simple, alternate, keeled or not, light green and sometimes white mottled, are quickly deciduous. The inflorescences are unusual in that they are cyathia, terminally located slipper-shaped, red bracts in the center of which is the female (pistillate) flower structure, surrounded by glandular structures and male (staminate) flowers (Figure 303). No petals are present. Fruits are 3-lobed capsules that split open to release the seeds.

Toxic Principle and Mechanism of Action

All parts of the plant contain irritating saponins that cause gastroenteritis if eaten in quantity.[1]

Risk Assessment

Pedilanthus species are often cultivated in xeriscape gardens, and can be used for hedges. In temperate climates the plant can be successfully grown indoors, and therefore can be a problem to the household pet that chews on the plants.

Clinical Signs

Excessive salivation, vomiting, and diarrhea may develop after the plant has been chewed and eaten. Symptoms are generally mild and transient. The sap may cause dermatitis in some people who are exposed to it.[2] Other plant species belonging to the Euphorbiaceae that have irritating sap that can cause similar signs include *Euphorbia tirucalli, E. myrsinites, Syadenium grantii* (African milk bush)

References

1. Burrows GE, Tyrl RJ: Toxic Plants of North America. Iowa State University Press, Ames 479-480, 2001.
2. Fuller TC, McClintock E. In Poisonous Plants of California. University of California Press, Berkley 140, 1986.

Persea americana

Family: Lauraceae

Common Names

Avocado, alligator pear, aguacate

Plant Description

A genus of about 150 species, *Persea* are native to tropical areas of the Americas and Asia, and are evergreen, branching, shrubs and trees growing to 20-30 m in height. Leaves are alternate, clustered near the ends of branches, large ovate to elliptic, with prominent central vein. Flowers are produced in panicles, each flower being greenish yellow.

Fruits are ovoid berries with a thick skin surrounding a greenish-white flesh that is edible when ripe. A single large seed is present in each fruit. The fruits of the Guatemalan variety of *Persea americana* generally have a warty, dark purple-black skin, while the Mexican variety has a smooth, green skinned fruit (Figures 304 and 305).

Figure 304 *Persea americana* leaves

Toxic Principle and Mechanism of Action

Only the Guatemalan (rough or warty skinned) varieties, and hybrids containing the Guatemalan cultivar such as "Fuerte" are known to be toxic to horses, goats, sheep, cattle, rabbits, mice, budgerigars, canaries, cockatiels, and ostriches fed either the leaves or the unripe fruits.[1-6] The active principle in avocado leaves responsible for the myocardial necrosis is persin.[7] Its precise action is yet to be determined. Goats appear to be particularly susceptible to poisoning, with as little as 30g/kg body weight of the leaves causing signs of congestive heart failure as a result of myocardial necrosis.

Figure 305 *Persea* fruits

Depending on the amount of the leaves consumed, myocardial degeneration with cardiac failure is the most common outcome. Goats fed 20g/kg body weight of the leaves have a marked decline in milk production due to a sterile mastitis caused by necrosis of the secretory epithelium of the mammary gland.[8] Horses consuming avocado leaves seem to develop edema of the head and neck.[5] Chickens and turkeys are relatively resistant to the toxic effects pf *Persea* species.

Risk Assessment

Avocado trees are commonly grown in tropical areas for their edible fruits, and as shady trees. The seeds are often germinated in the home as a children's project accomplished by placing the seed in water and observing its rapid growth. The seedling then can be planted outside in warm climates, or it can be grown as a potted houseplant. The seeds should also be considered toxic. *Persea americana* var. Guatemala and any hybrids should not be fed to caged birds or avocado plants placed in an aviary.

Clinical Signs

Severely poisoned animals usually present with respiratory difficulty and cardiovascular changes typical of congestive heart failure. Horses may have severe edema of the head, neck, and brisket areas, and show signs of colic and respiratory difficulty. In less acute poisoning, especially in lactating does, ewes, cows, and mares, agalactia, and non-infectious mastitis may be the most prominent signs.[7] Caged birds fed avocados may be found dead if not observed closely.

There is no specific treatment for affected animals, and those showing cardiovascular signs caused by myocardial degeneration have a poor prognosis. Treatment is generally supportive in nature, and aimed at relieving symptoms. Lactating animals with mastitis generally recover, although milk production remains markedly reduced.

On postmortem examination, pulmonary edema, hydrothorax, subcutaneous edema, and myocardial necrosis are the primary findings. Necrosis of the mammary glandular epithelium may be seen in lactating animals.[4-6,8,9]

References

1. Burger WP, Naude TW, Van Rensburg IB, Botha CJ, Pienaar AC. Cardiomyopathy in ostriches *(Struthio camelus)* due to avocado (*Persea americana* var. guatemalensis) intoxication. J South African Vet Assoc 65: 113-118, 1994.
2. Shropshire CM, Stauber E, Arai M. Evaluation of selected plants for acute toxicosis in budgerigars. J Am Vet Med Assoc 200: 936-939, 1992.
3. Stadler P, van Rensburg IB, Naude TW. Suspected avocado *(Persea americana)* poisoning in goats. J South African Vet Assoc 62: 186-188, 1991.
4. Grant R, Basson PA, Booker HH, Hofherr JB, Anthonissen M. Cardiomyopathy caused by avocado (*Persea americana* Mill) leaves. J South Afr Vet Assoc 62: 21-22, 1991.
5. McKenzie RA, Brown OP. Avocado *(Persea americana)* poisoning of horses. Austr Vet J 68: 77-78, 1991.
6. Hargis AM, Stauber E, Casteel S, Eitner D. Avocado *(Persea americana)* intoxication in caged birds.J Am Vet Med Assoc 194: 64-66, 1989.

7. Oelrichs PB, et al. Isolation and identification of a compound from avocado *(Persea americana)* leaves which causes necrosis of the acinar epithelium of the lactating mammary gland and the myocardium. Natural Toxins 3: 344-349, 1995.
8. Craigmill AL, Seawright AA, Mattila T, Frost AJ. Pathological changes in the mammary gland and biochemical changes in milk of the goat following oral dosing with leaf of the avocado *(Persea americana)*.Austr Vet J 66: 206-211, 1989.
9. Craigmill AL, Eide RN, Shultz TA, Hedrick K. Toxicity of avocado *(Persea americana* (Guatamalan var)) leaves: review and preliminary report. Vet Human Toxicol 26: 381-383, 1984.

Philodendron

Family: Araceae

Common Names
Philodendron, money plant, sweetheart vine.

Plant Description
This large and diversely complex genus of some 500 species is native to tropical America and the West Indies. In North America philodendrons are most frequently found as a house or garden plant. As a morphologically diverse genus, philodendrons can vary from small vine-like plants, to shrubs and large climbing species gaining considerable size in their natural tropical habitats. These evergreen plants have stout stems, often with and adventitious aerial roots that facilitate their climbing habit. Leaves are generally simple, dark green, leathery, heart or arrowhead-shaped, varying in size from a few inches to three to 4 ft. in length. In some species, the mature leaves become deeply load. In contrast to *Monstera deliciosa* (Ceriman), which is commonly mistaken for a philodendron, the leaves of philodendrons do not contain perforations or holes. Some species have red stems and purple-red new growth (Figures 306-308).

Figure 306 *Philodendron bipinnatifidum*

Figure 307 *Philodendron scanders* (variegated)

Philodendron scandens, a widely grown Mexican species with heart-shaped glossy green mature leaves, has immature foliage that is golden-brown. The fluorescence is a white, green, or yellow spathe often with red or purple margins overlapping to form tubes. The spadices are

Figure 308 *Philodendron* species

shorter than the spathes. Fruits are white or orange red berries. The common house and garden plant, pothos (*Epipremnum* species), is frequently mistaken for a philodendron.

Toxic Principle and Mechanism of Action

Like other members of the Araceae family, *Philodendron* species contain oxalate crystals in the stems and leaves.[1,2] The calcium oxalate crystals (raphides) are contained in specialized cells referred to as idioblasts.[1,2] Raphides are long needle-like crystals bunched together in these specialized cells, and when the plant tissue is chewed by an animal, the crystals are extruded into the mouth and mucous membranes of the unfortunate animal. The raphides once embedded in the mucous membranes of the mouth cause an intense irritation and inflammation. Evidence exists to suggest that the oxalate crystals act as a means for introducing other toxic compounds from the plant such as prostaglandins, histamine, and proteolytic enzymes that mediate the inflammatory response.[3] (see *Dieffenbachia* species)

In addition to the contact irritant effects of the philodendrons, the plants appear to have nephrotoxic and neurotoxic properties, especially in cats.[4-7] Signs of renal failure do not appear until several days after the ingestion of the plant.

Risk Assessment

Philodendrons are frequently grown for their foliage and hardiness as house plants and consequently household pets have access to the plants.

Clinical Signs

Household pets that chew repeatedly on the leaves and stems of philodendrons may salivate excessively and vomit as a result of the irritant effects of the calcium oxalate crystals embedded in their oral mucous membranes. The painful swelling in the mouth may prevent the animal from eating for several days. Severe conjunctivitis may result if plant juices are rubbed in the eye. If a cat is known to have ingested the leaves, it should be monitored for signs of renal failure for several days and treated accordingly.

Treatment

Unless salivation and vomiting are excessive, treatment is seldom necessary. Anti inflammatory therapy may be necessary in cases where inflammation and edema of the mouth is severe. The plants should be removed or made inaccessible to the animals eating them.

References
1. Genua JM, Hillson CJ: The occurrence, type, and location of calcium oxalate crystals in the leaves of 14 species of Araceae. Ann Bot 56: 351-361, 1985.
2. Franceschi VR, Horner HT: Calcium oxalate crystals in plants. Bot Rev 46: 361-427, 1980.
3. Saha BP, Hussain M: A study of the irritating principle of aroids. Indian J Agric Sci 53: 833-836, 1983.
4. Brogger JN: Renal failure from philodendron. Mod Vet Pract 51:46, 1970.
5. Pierce JH: Encephalitis signs from philodendron leaf. Mod Vet Pract 51: 42, 1970.
6. Sellers SJ, King M, Aronson CE, Der Maderosian A: Toxicologic assessment of *Philodendron oxycardium* Schott (Araceae) in domestic cats. Vet Hum Toxicol 20: 92-96, 1978..
7. Knight TE: Philodendron-induced dermatitis: report of cases and review of the literature. Cutis 48: 375-378, 1991.

Phoradendron
Family: Viscaceae (Loranthaceae)

Common Name
Mistletoe

Plant Description
Consisting of over 400 species in 11 genera, the mistletoes are distributed worldwide, and are parasitic plants whose roots penetrate the branches and trunks of trees as a means of acquiring nutrients. The leaves are simple, evergreen, leathery, and oval or scale-like depending upon the species. Plants can be monoecious or

Figure 309 Mistletoe berries

dioecious, with the rather inconspicuous, greenish-white flowers being produced in leaf axils. Fruits are one-seeded berries, the flesh being mucilaginous, and the color varying from white to pink or red depending upon the species (Figure 309).

The most common species of mistletoe found traditionally in households at Christmas time is *Phoradendron serotinum*, while the European counterpart is *Viscum album*.

Toxic Principle and Mechanism of Action
Numerous glycoprotein lectins are present in the leaves and fruits. The lectins inhibit protein synthesis in a similar manner to ricin and abrin, but are far less potent. Additionally, there are alkaloids found in the *Phoradendron* species that have vasoactive properties. Considerable research has been done to investigate the reported anticancer benefits of different mistletoe species.[1] The toxicity of mistletoe varies depending upon the species and may be associated with the host tree it is parasitizing.[2]

Risk Assessment

Historically the ingestion of mistletoe berries has been associated with poisoning and fatalities in humans. Mistletoe berries are one of the most frequently ingested berries found in people's homes.[3] However evaluation of mistletoe poisoning cases indicate that the ingestion of the berries has little or no effect and are rarely fatal unless eaten in large numbers.[4,5] It is reasonable to assume therefore, that household pets that eat only a few mistletoe berries will be minimally affected.

Clinical Signs

The most common signs of poisoning in humans include vomiting, abdominal pain, and diarrhea. Incoordination, hypotension, and cardiovascular collapse are occasionally reported. Treatment should be symptomatic and include fluids and electrolytes.

References

1. Mansky PJ: Mistletoe and cancer: controversies and perspectives. Seminars in Oncology. 29: 589-594, 2000.
2. Fukunaga T et al: Studies on the constituents of Japanese mistletoes from different host trees, and their antimicrobial and hypotensive properties. Chem Pharm Bull (Tokyo) 37: 1543-1546 1989.
3. Veltri JC, Litovitz: 1983 annual report of the American association of poison control centers national data collection system. Am J Emerg Med 2: 420-426, 1984.
4. Hall AH. Spoerke DG. Rumack BH: Assessing mistletoe toxicity. Annals Emergency Medicine. 15: 1320-1323, 1986.
5. Spiller HA, Willias DB, Gorman SE, Sanftleban L: Retrospective study of mistletoe ingestion. J Toxicol Clin Toxicol 34: 405-408, 1996.

Physalis

Family: Solanaceae

Figure 310 *Physalis lobata*

Common Names

Chinese or Japanese lantern – *Physalis alkekengi*
Cape gooseberry – *P. peruviana*
Tomatillo – *P. philadelphica*
Ground cherry – *P. lobata*

Plant Description

Comprising a genus of 80 or more species, *Physalis* are found world wide and are bushy, erect, annuals or perennials, some having a rhizomatous root system. Leaves are alternate or whorled, entire or pinnately lobed. Flowers are produced in the leaf axils, and are white, yellow, blue, or purple depending upon the species (Figures 310 and 311). The fruits are edible in some species and become enclosed by the calyces that are papery and are bright orange-red in some species.

Toxic Principle and Mechanism of Action

Solanine glycoalkaloids are found in all parts of the plants, with the ripe fruits having negligible amounts. Calystegins have been isolated from *P. alkekengi.*[1]

Risk Assessment

Cases of poisoning are rarely reported involving *Physalis*, but because certain species are grown for their edible fruits or colorful "lanterns", there is potential for the solanine alkaloids to cause toxicity in people or animals that eat the plants or unripe fruits.[2]

Figure 311 *Physalis alkekengi* Inset-Berry

Clinical Signs

Eaten in quantity, ground cherries can cause digestive problems, and in one report sheep died from liver disease and hemorrhagic lungs after eating cut-leaf ground cherry *(P. angulata).*[2]

References

1. Asano N, Kato A, Oseki K, Kizu H, Matsui K: Calystegins of *Physalis alkekengi* var. *francheti* (Solanaceae) structure determination and their glycosidase inhibitory activities. Eur J Biochem 229: 369-376, 1995.
2. Fuller TC, McClintock E: Poisonous Plants of California. Univ Calif Press, Berkeley pp 242-245, 1986.

Phytolacca

Family: Phytolaccaceae

Common Names

Pokeweed, pigeon berry, red ink plant - *Phytolacca americana* is native to the south Eastern area of North America. *P. decandra* is a South American species.

Plant Description

A genus of about 25 perennial species, *Phytolacca* are native to tropical and sub-tropical areas of the Americas, Africa, and Asia. Branching erect shrubs and small trees, with purple stems, alternate, ovate to

Figure 312 *Phytolacca americana*

lanceolate, and entire leaves. Inflorescences are racemes or panicles of small, cup-shaped, petalless flowers. Fruits are berries or drupes, turning purplish-black when ripe (Figures 312 and 313).

Figure 313 *Phytolacca americana* ripe fruits

Toxic Principle and Mechanism of Action

The roots in particular, but all parts of the plant contain the toxic saponins and oxalates. The triterpene saponins phytolactogenin and phytolaccinic acid and their respective glycosides are irritants especially to the gastrointestinal tract.[1,2] The oxalates may also contribute to the digestive irritation. Within the berries are mitogens that are well known for their ability to stimulate lymphocyte transformation.[3]

Risk Assessment

Pokeweed is occasionally planted as an ornamental, but more often is a weed whose seeds are spread by birds. The showy fruits and new stems and leaves are edible if properly prepared and cooked. However, the plant should be considered toxic as even cooking may still result in digestive problems.[4] Poisoning in animals has occurred when the plant has been eaten in quantity by cattle. The plant is also toxic to pigs and young turkey poults, and sheep have been fatally poisoned by the Brazilian species *P. decandra.*[5]

Clinical Signs

In cases of human poisoning, mouth irritation and burning, salivation, vomiting, and colic may be seen. Some people have developed diarrhea, sweating, tremors, cardiac irregularity, and hypotension leading to fainting.[4,6] Diarrhea and decreased milk production occurs in cattle eating large amounts of pokeweed.[7] Pigs eating the roots develop more severe signs including incoordination, seizures, paralysis, and death.[8] Turkey poults had decreased weight gains, ataxia, and swollen hocks.[9]

Treatment is generally directed at relieving the severest symptoms and should include intestinal protectants such as activated charcoal and intravenous fluids.

References

1. Johnston A, Schimuzu Y: Phytolaccinic acid, a new *Phytolacca americana.*Tetrahedron 30: 2033-2036, 1974.
2. Yi Y-H, Wang C-L: A new active saponin from *Phytolacca esculenta.* Plant Med 55: 551-552, 1989.
3. Waxdall MJ; Isolation, characterization and biological activities of 5 mitogens from pokeweed. Biochem 13: 3671-3676, 1974.
4. Callahan R et al: Plant poisoning- New Jersey. Morb Mortal Wkly Rep 30: 65-66. 1981.
5. Piexto P, Wouters F, Lemos RA, Loretti AP: *Phytolacca decandra* poisoning in sheep in Southern Brazil. Vet Hum Toxicol 39: 302-303, 1997.
6. Jaeckle KA, Freemon FR: Pokeweed poisoning. South Med J 74: 639-640, 1981.
7. Kingsbury JM, Hillman RB: Pokeweed *(Phytolacca)* poisoning in a dairy herd. Cornell Vet 55: 534-538, 1965.
8. Patterson FD: Pokeweed causes heavy losses in a swine herd. Vet Med 24: 114, 1929.
9. Barnett BD: Toxicity of pokeberries (fruit of *Phytolacca americana* Large) for turkey poults. Poultry Sci 54: 1215-1217, 1975.

Family: Ericaceae (Heath family)

Common Names
Japanese pieris, mountain pieris, mountain fetterbush, lily of the valley bush.

Plant Description
The 10 species of *Pieris* are native to the acidic, moist soils of the temperate regions of North America and eastern Asia. As small to large, erect, branching shrubs, leaves are ovate to lanceolate, alternate, glossy, reddish-bronze initially in the spring, turning dark green in maturity. Inflorescences are terminal panicles or racemes. The fragrant flowers are showy, white to dark pink, cylindrical or bell-shaped, pendent, with 5 fused sepals and 5 fused petals. Fruits are ovoid capsules (Figures 314 and 315).

Figure 314 *Pieris japonica* flowers

The species of *Pieris* most frequently associated with toxicity include *Pieris japonicus* (Japanese pieris) and *P. floribunda* (mountain pieris). Numerous cultivars exist, and are popular for their new foliage color and prolific flower production in the Spring.

Toxic Principle and Mechanism of Action
All species of the family Ericaceae contain varying quantities of toxic diterpenoids collectively known as grayanotoxins I and II (formerly andromedotoxin, rhodotoxin, and acetylandromedol).[1] As many as 18

Figure 315 *Pieris japonica* 'pink'

grayanotoxins (I – XVIII) have been identified, the greatest number being found in the *Leucothoe* species (fetter bush).[2,3] Tannins and other compounds are also present in varying amounts. All parts of the laurel including the flowers are toxic, although there may be considerable variation between species.

Grayanotoxins act to increase sodium channel permeability of cells by opening the channels to sodium, which enters the cells in exchange for calcium ions, thus rendering the channels slow to close so that the cell remains depolarized.[4,5] Other neurologic mechanisms may also involve a cholinergic response seen clinically as bradycardia and excessive salivation.[6] The cardiac effects can range from bradycardia, sinus arrest, and arrhythmias.

Other members of the Ericaceae that contain grayanotoxins include:

Andromeda polifolia	Andromeda, bog rosemary
Kalmia spp.	Laurel
Ledum spp.	Labrador tea
Leucothoe spp.	Fetter bush, dog laurel
Lyonia spp.	Maleberry
Menziesia spp.	Rusty menziesia
Rhododendron spp.	Rhododendrons, azaleas

Risk Assessment

Pieris species are commonly grown as showy garden shrubs. Livestock poisoning occurs where the plants are accessible to the animals.

Clinical Signs

Excessive salivation, increased nasal secretions, vomiting, abdominal pain, bloat, and irregular respirations develop several hours after *Pieris* leaves are ingested.[8,9] Projectile vomiting may be noticeable. Hypotension, tachycardia, and respiratory depression may also develop. Weakness, partial blindness, and seizures have been reported in severe intoxications. Neurologic signs may persist for several days before the animal recovers. Weight loss may be notable. Death may result from the inhalation of vomited rumen contents.[10] Pregnant goats eating *Pieris* can cause fetal mummification.[11]

Treatment is primarily directed at relief of the more severe clinical signs. Activated charcoal given orally is helpful if given shortly after the *Pieris* is consumed. Atropine is useful in countering the cardiovascular effects.

References

1. Kakisawa H, Kozima T, Yanai M, Nakanishi K: Stereochemistry of grayanotoxins. Tetrahedron 21: 3091-3104, 1965.
2. Sakikabara J, Shirai N, Kaiya T, Nakata H: Grayanotoxin-XVIII and its grayanoside B, a new A-Nor-B-Homo-Ent-Kaurine and its glucoside from *Leucothoe grayana*. Phytochemistry 18: 135-137, 2979.
3. Sakikabara J, Shirai N, Kaiya T: Diterpene glycosides from *Pieris japonica*. Phytochemistry 20: 1744-1745, 1981.
4. Narahashi T, Seyama I: Mechanism of nerve membrane depolarization caused by grayanotoxin I. J Physiol 242: 471-487, 1974.
5. Seyama I, Narahashi T: Modulation of sodium channels of squid nerve membranes by grayanotoxin I. J Pharmacol Exp Ther 219: 614-624, 1981.
6. Onat F, Yegan BC, Lawrence R, Oktay A, Oktay S: Site of action of grayanotoxins in mad honey in rats. J Appl Toxicol 11: 199-201, 1991.
7. Krochmal C: Poison honeys. Am Bee J 134: 549-550, 1994.
8. Casteel SW, Wagstaff JD: *Rhododendron macrophyllum* poisoning in a group of goats and sheep. Vet Hum Toxicol 31: 176-178, 1989.
9. Puschner B et al: Azalea (*Rhododendron* spp.) Toxicosis in a group of Nubian goats. Proc 42nd Annual Meeting Am Assoc Vet Lab Diagn. 1999.
10. Smith MC: Japanese pieris poisoning in the goat. J Am Vet Med Assoc 173: 78-79, 1978.
11. Smith MC: Fetal mummification in a goat due to Japanese pieris (Pieris japonica) poisoning. Cornell Vet 69: 85-87, 1979.

Pistia stratiotes

Family: Araceae

Common Names
Water lettuce, floating lettuce, shell flower

Plant Description
A monotypic genus, *Pistia stratiotes*, is a widespread aquatic perennial of the tropics, and has become a noxious weed in many lakes and waterways. With 6 inch (15 cm) leaves arranged in a rosette like a lettuce head, the plant floats on water with the assistance of the spongy air-filled base of the leaves. The plant reproduces via rosettes that connect to the parent plant. The arum-like flowers are inconspicuous, as they are enclosed in a leaflike spathe (Figure 316).

Figure 316 *Pistia stratiotes*

Toxic Principle and Mechanism of Action
Pistia startiotes contains oxalate crystals in the stems and leaves. The calcium oxalate crystals (raphides) are held in specialized cells referred to as idioblasts.[1,2] Raphides are long needle-like crystals grouped together in the specialized cells, and when the plant tissue is chewed by an animal, the crystals penetrate the mouth and mucous membranes of the unfortunate animal. The raphides once embedded in the mucous membranes cause an intense irritation and inflammation.

Risk Assessment
Water lettuces are common in tropical ponds, lakes and waterways, and are frequently grown in garden ponds and aquariums to help cover the water surface, thereby reducing the water temperature and the growth of algae. Dogs playing in the water and chewing the vegetation are susceptible to the irritant effects of the calcium oxalate crystals in the leaves.

Clinical Signs
Excessive salivation and vomiting are the most likely encountered clinical signs. Treatment is seldom necessary as the effects tend to be transitory.

References
1. Franceschi VR, Horner HT: Calcium oxalate crystals in plants. Bot Rev 46: 361-427, 1980.
2. Genua JM, Hillson CJ: The occurrence, type, and location of calcium oxalate crystals in the leaves of 14 species of Araceae. Ann Bot 56: 351-361, 1985.

Pittosporum
Family: Pittosporaceae

Figure 317 *Pittosporum tobira* 'variegatum'

Common Names
Pittosporum, Japanese mock orange, lemonwood, Victorian box.

Plant Description
Consisting of some 200 species of evergreen shrubs or small trees from the tropical and subtropical regions on Australia, Asia, Africa, and the Pacific Islands, *Pittosporum* species are widely grown in tropical and subtropical areas of North America. The leaves are simple, alternate and arranged at the ends of branches. Some species have variegated leaves. The fragrant flowers are produced terminally or in leaf axils, as panicles or clusters, with white, yellow, pink, or red petals. Fruits consist of a woody capsule containing sticky seeds (Figure 317).

Toxic Principle and Mechanism of Action
The leaves and fruits contain saponins that when ingested cause gastrointestinal irritation.

Risk Assessment
Since *Pittosporum* species and their cultivars are commonly grown as garden or houseplants, the fruits and leaves are a potential hazard to dogs and cats that may chew on them.

Clinical Signs
Excessive salivation, abdominal pain, and vomiting can be anticipated with *Pittopsorum* poisoning.

References
1. Burrows GE, Tyrl RJ: Toxic Plants of North America. Iowa State University Press, Ames pp 1238, 2001.

Plumbago

Family: Plumbaginaceae

Common Names
Leadwort, plumbago

Plant Description
A genus of about 15 species that are native to many warm-temperate and tropical areas of the world. These evergreen, annual, or perennial shrubs have alternate, simple, leaves. Inflorescences are produced terminally, in racemes. Flowers are showy, radially symmetrical, 5-petalled, and trumpet-shaped (Phlox-like). The fruits are capsules. The commonest species in cultivation is *P. auriculata* (Figures 318 and 319).

Figure 318 *Plumbago auriculata*

Toxic Principle and Mechanism of Action
All parts of the plant contain the irritant naphthaquinone plumbaginin which has a variety of effects on the reproductive system including abortion, irregular estrus cycles, and testicular degeneration.[1,2] Some species of *Plumbago* cause gastroenteritis and diarrhea, and the Brazilian species *P scandens* cause deaths in cattle and goats.[3,4]

Figure 319 *Plumbago auriculata* flowers

Risk Assessment
Plumago is unlikely to be a problem to pets, but it is a common garden shrub in warm-temperate and tropical areas.

Clinical Signs
Salivation, vomiting, and diarrhea are likely to be the primary clinical signs encountered.

References
1. Thompson RH. Naturally occurring quinines, 2nd ed. Academic Press, London 1971.
2. Bhargava SK. Effects of plumbagin on the reproductive function of the male dog. Indian J Exp Biol 1984, 22: 153-156.
3. Tokarnia CH, Dobereiner J. Experimental poisoning of cattle with *Plumbago scandens* (Plumbaginaceae). Pesq Vet Bras 1982, 2: 105-112.
4. Medieros RMT, Barbosa RC, Lima EF, Simoes SVD. Intoxication of goats by *Plumbao scandens* in Northeastern Brazil. Vet Hum Toxicol 2001, 43: 167-168.

Family: Apocynaceae

Figure 320 *Plumeria obtusa*

Figure 321 *Plumeria obtusa* flowers

Common Names
Plumeria, frangipani, pagoda tree, temple tree, nosegay, graveyard tree, West Indian jasmine, Singapore plumeria.

Plant Description
A genus of 8 species native to the tropical and subtropical Americas, *Plumeria* species are perennial, deciduous, branching shrubs or small trees with cylindrical stems, obvious leaf scars, and a milky sap. Leaves are leathery, alternate, or spirally arranged, clustered at the ends of branches with long petioles. Leaves are glossy, dark green, with conspicuous veins, the secondary veins being joined by a nerve running parallel to the leaf margin. Inflorescences are showy, terminal cymes or panicles often appearing before the leaves. Flowers are fragrant, with 5 broad petals arranged at right angles to a slender tube, and colors range from white with a yellow center to yellow to shades of pink or red. Fruits are paired, long, tubular follicles containing numerous winged seeds (Figures 320, 321, and 322).

Toxic Principle and Mechanism of Action
A variety of iridoids are found in the bark of the plants.[1,2] These compounds are gastrointestinal irritants causing diarrhea and anorexia, and can be embryotoxic.[3] Also present in the plant are a variety of pentacyclic terpenoids that have some cardiotoxic effects.[4] The sap is irritating to the skin and will cause mouth irritation if the plant is chewed.

Risk Assessment
Plumeria species are common garden plants in tropical areas, grown especially for the showy fragrant flowers and attractive growth form. The plants do well in green houses and as house plants. Cats and dogs chewing the stems and leaves may experience the irritant effects of the sap.

Clinical Signs

Excessive salivation, vomiting, and diarrhea may be expected if household pets eat the plants. Rarely are the symptoms severe and animals soon recover without the need for treatment.

Figure 322 *Plumeria rubra*

References

1. Kardono LBS, Tsauri S, Padmawinata K, Pezzuto JM, Kinghorn AD: Cytoxic constituents of the bark of *Plumeria rubra* collected in Indonesia. J Nat Prod 53: 1447-1455, 1990.
2. Connolly JD, Hill RA: Dictionary of Terpenoids. Vol 1, Mono Terpenoids. Chapman & Hall, London 1991a.
3. Gunawardana VK, Goonesekera MM, Gunaherath GMKB: Embryotoxic effects of *Plumeria rubra*. Toxic Plants and Other Natural Toxins, Garland T, Barr AC eds. New York 317-322, 1998.
4. Radford DJ, Gillies AD, Hinds JA, Duffy P. Naturally occurring cardiac glycosides. Medical J Australia. 144: 540-544, 1986.

Podocarpus

Family: Podocarpaceae

Common Names

Buddhist pine, yew pine, kusamaki

Plant Description

A genus of about 100 species of coniferous evergreen ground covers to large trees 45m in height, *Podocarpus* are native to tropical areas of the southern hemisphere and extending into Mexico and Japan. *Podocarpus macrophylla* is a cold tolerant species from Japan where it matures to a tree up to 21 m in height. Leaves are long,

Figure 323 *Podocarpus macrophylla*

linear, leathery, glossy green and spirally arranged. Male and female flowers are produced on separate plants. The male flowers are catkin-like pollen cones, while the female flowers have seeds held on slender stalks that develop into fleshy, berry-like fruits that are black or red in color. The leaves are similar in appearance to those of yews (*Taxus* species) but much larger (Figures 323 and 324).

Toxic Principle and Mechanism of Action

The toxin in the fruits and leaves of the plants that causes vomiting and diarrhea has not been determined.

Figure 324 *Podocarpus* fruits

Risk Assessment
Commonly grown as a specimen tree, hedge, or as a potted indoor plant, *Podocarpus* species have potential for causing poisoning if the berries are consumed.

Clinical Signs
The National Animal Control Center lists *Podocarpus* as toxic and capable of causing severe vomiting and diarrhea.[1] Treatment when necessary should be directed at relieving the gastrointestinal symptoms.

References
1. Household Plant Reference. ASPCA National Animal Poison Control Center. 424 E. 92nd St., New York, NY 10128

Podophyllum
Family: Berberidaceae

Figure 325 *Podophyllum peltatum*

Common Names
May apple, umbrella leaf, American mandrake, Indian apple, witch's umbrella

Plant Description
Only one species *(Podophyllum peltatum)* of some 7 species found in North America and Europe occurs in North America. As a common wildflower of eastern North America, it is often grown in wooded areas for its spreading habit as a ground cover. The leaves emerge before the foliage of deciduous trees. Glossy green palmate, leaves are produced on stems 18-24 inches tall, in pairs, with 5-7 lobes or partitions. Leaf veins are prominent and palmately arranged. A single creamy-white, 5-7 petalled, nodding flower arisng at the angle between the petioles is produced in early spring. The canopy of leaves hide the blooms. The fruit is an oval berry containing many seeds that turns yellow when ripe (Figures 325 and 326).

Toxic Principle and Mechanism of Action
The primary toxin in *Podophyllum* species is podopyllotoxin, one of many lignan genins and glycosides found in greatest concentration in the plant at the time of flowering.[1] The ripe fruit has little toxicity. Podophyllotoxins are readily absorbed through intact skin and the digestive tract. The toxicity of podophyllotoxin is

attributed to its binding to receptor sites on tubulin, thereby blocking cell division and cellular protein synthesis in a similar manner to colchicine found in the autumn crocus. Podopyllotoxin has been extensively utilized in folk medicine and currently has value for its antiviral and anti-neoplastic properties.[1,2]

Risk Assessment
May apple is often grown as a woodland ground cover in eastern North America where it appears in early spring before deciduous trees produce their leaves. The risk of poisoning of family pets is low, but caution is warranted.

Figure 326 *Podophyllum peltatum* fruit

Clinical Signs
Animals or humans eating the green "apples" are likely to experience severe diarrhea and abdominal pain. Excessive salivation and vomiting may be prominent. Ingestion of large amounts of the podophyllotoxin can lead to neurologic signs, liver degeneration, and bone marrow dysfunction.[3] Poisoning in humans has occurred when *Podophyllum* products have been mistaken for mandrake *(Mandragora officiinarum)*, an herbal medication with quite different effects.[4]

References
1. Canel C Moraes RM, Dayan FE, Ferriera D: Podophyllotoxin. Phytochemistry 54: 115-120, 2000.
2. Damayanthi Y, Lown LW: Podophyllotoxins: current status and recent developments. Current Med Chem 5: 205-252, 1998.
3. Filley CM et al: Neurologic manifestation of podophyllin toxicity. Neurology 32: 308-311, 1982.
4. Frasca T, Brett AS, Yoo SD: Mandrake toxicity: a case of mistaken identity. Arch Int Med 157: 2007-2009, 1997.

Portulaca

Family: Portulacaceae

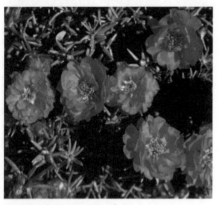

Figure 327 *Portulaca* species

Common Names
Portulacca, moss rose, rock moss, sun plant – *Portulaca grandiflora*
Purslane – *P. oleraceae*

Plant Description
Comprising a genus of 100 semi-succulent, erect or prostrate, mostly perennials, *Portulaca* are native to warm drier regions of the world. Leaves are alternate or opposite, fleshy, round to flattened, the upper few leaves forming an involucre surrounding the showy flowers, with 4-6 petals and prominent multiple stamens. Flowers have a wide range of colors ranging through white, yellow, pink, purple, or red and variously striped (Figure 327).

Toxic Principle and Mechanism of Action
Oxalates are the primary toxins in *Portulaca*. The common garden weed purslane *(P. oleraceae)* may have oxalate levels as high as 17% dry matter.[1] Purslane has been used as a food source as long as it is well cooked to remove the soluble oxalates. Animals have been poisoned by eating large quantities of the plant and developed typical signs of oxalate poisoning seen with other oxalate containing plants such as wood sorrel (*Oxalis* spp.)[1] If eaten in quantity the oxalates can induce hypocalcemia and oxalate nephrosis.

Risk Assessment
Portulaca are commonly grown as summer time drought tolerant plants in many gardens. The common weed purslane *(Portulaca oleraceae)* is frequently removed in large quantities from gardens and may be disposed of in animal enclosures where it may be readily consumed by livestock. In quantity, the plants can cause fatal poisoning.

Clinical Signs
Sheep that have eaten large amounts of the plant become depressed, ataxic, and recumbent because of hypocalcemia following the formation of insoluble calcium oxalate. Oxalate crystals may be seen in the urine and animals die from an oxalate nephrosis.

In the event oxalate poisoning is confirmed, the intravenous administration of calcium borogluconate may rapidly reverse hypocalcemic effects, but oxalate nephrosis may not be affected and animals may die several days later from renal failure.

References
1. Libert B, Franceschi VR: Oxalate in crop plants. J Agric Food Chem 35: 926-938, 1987.

Prunus

Family: Rosaceae

Common Names
Choke cherry *(Prunus virginiana)*, black cherry *(P. serotina)*, cherry laurel *(P. laurocerasus)*, and service berry *(Amelanchier* species) are of toxicologic significance.

Plant Description
A diverse genus of some 400 species of woody, deciduous shrubs and small trees, *Prunus* are native to many areas of the Northern Hemisphere. Many species have thorns and smooth bark with obvious lenticels. Leaves are simple, alternate, petiolate, entire or serrate, and glossy green above. Inflorescences are racemes or clusters produced terminally on branches or woody spurs. Flowers have 5 sepals and 5 petals, in white, pink, or red. Fruits are fleshy drupes, each with a hard endocarp or stone, and one round or flattened seed per stone (Figures 328 and 329).

Figure 328 *Prunus* flowers

Figure 329 *Prunus* virginiana-choke cherries

Toxic Principle and Mechanism of Action
Many *Prunus* species contain cyanogenic glycosides, the primary glycosides being amygdalin (Laetrile) and prunasin.[1] Prunasin is especially high in the new growth and in wilted or frosted leaves, while amygdalin is highest in the seeds. The ripe pulp of the berries surrounding the seeds is not toxic.

At least 2000 plant species, many from the 27 genera represented in the Rosaceae contain cyanogenic glycosides, but relatively few have been associated with poisoning in livestock, and fewer still have affected household pets. The major sources of cyanogenic glycosides responsible for poisoning of ruminants are sorghums such as Johnson and Sudan grasses (*Sorghum* species), and a number of the *Prunus genus*. Ruminants are particularly sensitive to plant-induced cyanide poisoning because the cyanogenic glycosides are rapidly hydrolysed in the rumen by plant and bacterial enzyme action to produce the highly toxic hydrogen cyanide (HCN, or prussic acid).[2] Simple stomached animals such as horses, pigs, and dogs are rarely poisoned by plant cyanogenic glycosides because their digestive systems lack the ability to rapidly hydrolyse the glycosides to HCN. Dogs are very rarely affected by cyanide poisoning from plant origin. There are a few occasions where dogs have eaten quantities of choke cherry leaves and berries and heavenly-bamboo *(Nandina domestica)* and developed cyanide poisoning.[3,4]

Dogs have also been experimentally poisoned with fresh sweet almonds *(P. dulcis)* containing amygdalin.[5] Since apricot seeds *(P. armeniaca)* are also high in amygdalin,[6] dogs that eat quantities of fallen apricots have the potential for poisoning, especially if the pits are cracked open to expose the seeds to the animals digestive system.

It should be emphasized however that all animals are highly susceptible to HCN.[2] Dogs and cats are quite frequently poisoned by HCN produced in house fires and from eating or drinking chemicals containing sodium or potassium cyanide. It is because plants do not contain free HCN, and it requires enzymatic hydrolysis of the cyanogenic glycosides for the plant to become toxic, that dogs and other simple stomached animals are rarely affected by cyanogenic plants. Ruminants are most susceptible to plant induced cyanide poisoning.

Hydrogen cyanide is rapidly absorbed from the respiratory and digestive tracts and acts primarily by blocking mitochondrial cytochrome oxidase, the critical enzyme essential for oxygen ion transport for cellular energy needs.[2,7] Because the brain has the highest energy needs, HCN rapidly affects brain function, causing anoxia and death.

Risk Assessment
Plant induced cyanide poisoning in dogs and cats is rarely encountered. None the less, it is a wise precaution to avoid exposing household pets unnecessarily to those plants that are known to have high cyanogenic glycoside potential such as choke cherries, service berries, almonds, and apricots. Such plants should not be planted in dog runs where the fruits may fall and be eaten by the dogs.

Other garden plants of the family Rosaceae with cyanogenic glycoside potential include cotoneaster *(Cotoneaster* spp.), Christmas berry *(Heteromeles arbutifolia),* crab apples *(Malus* spp.), Photinia, Japanese photinia *(Photinia* spp.), firethorn, pyracantha *(Pyracantha coccinea),* and Spirea *(Spirea* spp.).[1]

Clinical Signs
Due to the peracute effects of HCN on the oxygen transport system in animals, it is not unusual for the animal to be found in acute respiratory distress or dead within about 30 minutes of ingesting a toxic dose of the plant. Initially mucous membranes are bright red in color due to oxygen saturation of hemoglobin. As the animal becomes severely hypoxic the mucous membranes may appear cyanotic before coma and death occur.

Animals suspected of cyanide poisoning require immediate treatment with as little stress placed on the animal as possible. Intravenous administration of sodium nitrite and sodium thiosulfate is the treatment of choice.[8,9] The nitrite has a great affinity for hemoglobin and forms cyanmethemoglobin thereby removing the cyanide from the cytochrome oxidase.[7,9] Sodium thiosulfate then binds with the cyanide ion to produce thiocyanate that can be excreted by the kidneys. A 20% solution of sodium nitrite should be administered slowly intravenous at a dose of

16mg/kg body weight, followed by a 20% solution of sodium thiosulfate at a dose of 1.65ml/kg body weight.[10] A solution of sodium thiosulfate given via stomach tube may help to reduce further absorption of HCN from the stomach but it has no effect on HCN already absorbed. Hydroxocobalamin, and alpha-ketoglutaric acid hav been shown to be an effective adjunct to treating cyanide poisoning in dogs.[11,12]

References
1. Burrows GE, Tyrl RJ: Toxic Plants of North America. Iowa State University Press, Ames. 1043-1074, 2001.
2. Salkowski AA, Penney DG: Cyanide poisoning in animals and humans: a review. Vet Hum Toxicol 36: 455-466, 1994.
3. Bradley M, Nieman LJ, Burrows GE: Seizures in a puppy. Vet Hum Toxicol 30: 121, 1988.
4. Stauffer VD: Hydrocyanic acid poisoning from choke cherry leaves. J Am Vet Med Assoc 157: 1324, 1970.
5. Schmidt ES, Newton GW, Sanders SM, Lewis JP, Conn EE: Laetrile toxicity studies in dogs. J Am Med Assoc 239: 943-947, 1978.
6. Suchard JR, Wallace KL, Gerkin RD: Acute cyanide toxicity caused by apricot kernel ingestion. Annals Emerg Med. 32: 742-744, 1998.
7. Osweiler GD: Toxicology. Williams & Wilkins. Philadelphia 400-401, 1996.
8. Burrows GE: Cyanide intoxication in sheep: therapeutics. Vet Hum Toxicol 23: 22-28, 1981.
9. Pickrell JA, Oehme F: Cyanogenic glycosides. In Clinical Veterinary Toxicology. Plumlee KH. Mosby, St Louis, Missouri. 2004, 391-392.
10. Gfeller RW, Messonier SP: handbook of small animal toxicology and poisonings. 2nd ed. Mosby, St Louis, Missouri 2004, 141-143.
11. de La Coussaye JE. Houeto P. Sandouk P. Levillain P. Sassine A. Riou B. Pharmacokinetics of hydroxocobalamin in dogs. J Neurosurg Anesth 1994, 6: 111-115.
12. Dalvi RR, Sawant SG, Terse PS: Efficacy of alpha-ketoglutaric acid as an effective antidote in cyanide poisoning in dogs. Vet Res Communications. 1990, 14: 411-414.

Family: Rosaceae

Figure 330 *Pyracantha coccinea*

Figure 331 *Pyracantha coccinea* fruits

Common Name
Pyracantha, firethorn – *Pyracantha coccinea*
Numerous cultivars exist.

Plant Description
A genus of 7 species from Southern Europe and Asia, Pyracantha are evergreen, erect or spreading woody shrubs or small trees with prominent thorns. Leaves are dark glossy green, alternate, simple and elliptic. Inflorescences are corymbs produced terminally. Flowers are hawthorn-like, showy white, with 5 sepals and 5 petals. Fruits are round to spherical berries in colors ranging from yellow to orange-red (Figures 330 and 331).

Toxic Principle and Mechanism of Action
The cyanogenic glycoside prunasin has been identified.[1] It is unlikely that cyanide poisoning would result from this plant.

Risk Assessment
Pyracanthas are commonly grown as garden and landscaping plants for their display of spring flowers and the masses of orange-red berries in late summer and fall. Poisoning is rarely encountered, although the large thorns in some species can cause injury. Birds, especially robins, have been observed by the author to have difficulty hopping and flying erratically after feeding on the fermenting berries in late summer.

References
1. Burrows GE, Tyrl RJ. Toxic Plants of North America. Iowa State University Press, Ames. 1065, 2001.

Quercus

Family: Fagaceae

Common Name
Oak

Plant Description
A genus of some 600 species of long-lived shrubs to large trees from temperate and tropical areas, oaks are perennials and mostly deciduous, although oaks of tropical and sub-tropical areas are evergreen, or "live-oaks". Leaves are usually glossy green, and vary considerably in shape from characteristic lobed leaves to some species that have entire or serrate edged leaves. Staminate (male) flowers are produced as yellowish catkins, while the female or pistillate flowers are greenish and inconspicuous The fruits are nuts (acorns) with a characteristic scaly, basal, cup-like carp. As a general rule white oaks, those with rounded leaf lobes have edible acorns, while red oaks with pointed leaf lobes tend to be bitter and inedible (Figures 332 and 333).

Figure 332 *Quercus gambelii*

Figure 333 *Quercus rubra* - Fall color

Toxic Principle and Mechanism of Action
The leaves in particular contain polyphenolic compounds such as tannic acid, and gallic acid that are astringents, or compounds capable of precipitating protein.[1] When absorbed in large quantity the tannic acid reacts with the protein of cells, denaturing it, and in the process destroys the cells. The primary sites of tannic acid damage are in the intestines, liver, and kidneys. There is great variability amongst animal species as to their susceptibility to oak poisoning. Goats and deer can thrive while browsing on oak, and even cattle can eat oak provided it does not become a predominant component of their diet.

Risk Assessment
Oak leaves pose the greatest risk to livestock that eat them when there are no other forages available. All animals are susceptible to oak poisoning depending on the quantity eaten.[2,3] Large quantities of acorns that have fallen to the ground may be a problem to horses that eat them in quantity. Occasionally avian species may be poisoned if they eat quantities of the leaves or acorns.[4] No reports incriminate oaks a cause of toxicity in dogs or cats. Consumption of acorns by dogs however has the potential for causing poisoning as occurs in other animals.

Clinical Signs

Initially anorexia and constipation followed by diarrhea and colic can be expected. This is followed by dehydration and renal failure with proteinuria, glucosuria hematuria, and marked elevations of blood urea nitrogen and creatinine levels.[5] The prognosis is poor once renal failure has developed as renal tubular degeneration is severe.

Treatment

Intravenous fluid and electrolyte therapy is indicated to help maintain kidney function. However, the animal should be carefully monitored as vascular damage may result in the leakage of fluids into body compartments compromising organ function. Intestinal protectants may be helpful in reducing the gastrointestinal consequences of oak poisoning

References

1. Adzet T, Camarasa J: Pharmocokinetics of polyphenolic compounds. Recent Adv Bot Hortic Pharmacol 3: 25-47, 1988.
2. Duncan CS: Oak leaf poisoning in two horses. Cornell Vet. 51: 159-62, 1961.
3. Garg SK et al: Oak *(Quercus incana)* leaf poisoning in cattle.Vet Hum Toxicol. 34: 161-164, 1992.
4. Kinde H: A fatal case of oak poisoning in a double-wattled cassowary *(Casuarius casuarius)*. Avian Dis. 1988, 32: 849-851.
5. Spier SJ et al: Oak toxicosis in cattle in northern California: clinical and pathologic findings. J AM Vet Med Assoc 1987, 191: 958-964.

Ranunculus

Family: Ranunculaceae

Figure 334 *Ranunculus acris*

Common Names

Buttercups, crow foot, butter cress, figwort

Plant Description

Consisting of at least 400 species native to most temperate regions of the world, buttercups can be both weeds or colorful wildflowers that have been hybridized into showy ornamentals. The Asian species *(Ranunculus asiaticus)* have been selectively hybridized for their double flowering habit. Many species of buttercup are common in meadows and in marshy areas as wild flowers, while some such as the bur buttercup *(Ranunculus (Ceratocephalus) testiculatus)* are invasive weeds and have caused poisoning in sheep.[1] Over 70 species of native or introduced species of buttercup occur in North America.

Ranunculus species are annuals or perennials growing from tuberous roots, rhizomes, or stolons. The leaves are alternate, palmate, simple or compound, margins entire or toothed, basal leaves having long petioles. Flowers can be small or showy, sepals 3-5, petals may be absent or as many as 25, color is often yellow, but can be white, red, or green (Figures 334 and 335).

Figure 335 *Ranunculus asiaticus* hybrid

Toxic Principle and Mechanism of Action
Ranunculus species contain the irritant glycoside ranunculin that is converted to protoanemonin when the plant tissues are chewed and macerated.[2] Protoanemonin is the vesicant, and it is polymerized to the non-toxic anemonin. The dried plant contains mostly anemonin and is therefore not toxic. Buttercups are most toxic when flowering containing 1-2% protoanemonin on a dry weight basis.[1]

Risk Assessment
Buttercups are not a significant problem to household pets as the bitter irritant effects of the plants are a deterrent to most dogs and cats. However, the showy "Ranunculus" that are sold as potted plants or as garden ornamentals have the potential to be chewed and eaten by pets.

Clinical Signs
Excessive salivation, vomiting, and diarrhea can be anticipated if buttercups are eaten in quantity.
Treatment if necessary should include activated charcoal (2-8 gm/kg body weight orally) and a cathartic such as magnesium sulfate to help in the removal of the plant material from the gastrointestinal tract. Intrvenous fluid therapy is indicated in the dehydrated animal.

References
1. Nachman RJ, Olsen JD: Ranunculin: a toxic constituent of the poisonous range plant bur buttercup *(Ceratocephalus testiculatus)*. J Agric Food Chem 31: 1358-1360, 1983.
2. Hill R, Van Heyningen R: Ranunculin: the precursor of the vesicant substance I Buttercup. Biochem J 49: 332-335, 1951.

Family: Rhamnaceae

Figure 336 *Rhamnus cathartica*

Common Name
Buckthorn
Common species include *R. californica* (California buckthorn, coffee berry, pigeon berry), *R. caroliniana* (Carolina buckthorn, Indian cherry), *R. cathartica* (common buckthorn), *R. frangula* (alder buckthorn).

Plant Description
A genus of about 125 species of deciduous or evergreen usually thorny shrubs to small trees, *Rhamnus* are native to many temperate regions of the world. Erect, branching, 1-10m tall shrubs or trees, with alternate, petiolate, ovate to oblong leaves, some variegated.

Inflorescences are umbels or solitary flowers produced in leaf axils. Flowers are small, greenish to yellow-white, sepals 4-5, triangular, petals 4-5 or absent. Fruits are berry-like drupes that can be brown, orange, or black in color (Figure 336).

A shrub or small tree native to southwestern North America commonly referred to as buckthorn, coyotillo and wild cherry to mention but a few of many common names is *Karwinskia humboldtiana*. The ripe fruits are edible but the seeds and foliage are poisonous to people and animals. Toxic anthracenones in the seeds cause degeneration of the peripheral nerves.[1] *Karwinskia* is rarely grown as an ornamental and is more of a problem to livestock browsing on the plants in arid areas and to children who may eat the unripe fruits and seeds.[2]

Toxic Principle and Mechanism of Action
The purgative action is due to anthrones and anthraquinones in the fruits and to a much lesser extent in the leaves.[3] *Rhamnus purschiana* is the origin of cascara sagrada, a well known cathartic.

Risk Assessment
Buckthorns are commonly grown for their attractive leaves, blossoms that attract bees, and for their fruits. Consequently children and pets have the potential for poisoning from the berries. The leaves are minimally toxic if eaten in very large quantities.

Clinical Signs
Vomiting, abdominal pain, and diarrhea that can become hemorrhagic can be expected.[2] In severe cases dehydration may result and will require fluid therapy. Activated charcoal and other adsorbents should be given orally.

References
1. Calderon-Gonzalez R, Rizzi-Hernadez H: Buckthorn polyneuropathy. N Eng J Med 277: 69-71, 1967.
2. Burrows GE, Tyrl RJ: Toxic Plants of North America. Iowa State University Press, Ames. 1034-1042, 2001.
3. Thompson RH: Naturally occurring quinines. 2nd Ed. Academic Press, London. 1971.
4. Burrows GE, Tyrl RJ. Toxic Plants of North America. Iowa State University Press, Ames. 1038-1042, 2001.

Rhododendron
Family: Ericaceae (Heath family)

Common Names
Rhododendron, azalea, mountain rosebay, red laurel, rosebay laurel, great laurel, and many regional names such as Cascade rhododendron, California rosebay

Plant Description
With as many as a 1000 species, and as many or more cultivars, rhododendrons are native to the acidic soils of the temperate northern hemisphere and southeast Asia. In general, the name azalea is given to the deciduous species, while the evergreen species are called rhododendrons.

Figure 337 *Rhododendron* (Azalea)

Ranging from small shrubs to erect, branching small trees with rough or smooth bark, *Rhododendron* species have alternate, glabrous or hairy, dark green, revolute, deciduous or evergreen, elliptic to lanceolate leaves. Inflorescences are terminal umbellate, racemose or corymbose clusters of showy flowers. Consisting of 5 fused sepals and 5 fused petals, the flowers are generally funnel-shaped, in virtually all colors except blue (Figures 337-340). Fruits are hard or soft capsules.

Figure 338 *Rhododendron* species

Toxic Principle and Mechanism of Action
All species of the family Ericaceae contain varying quantities of toxic diterpenoids collectively known as grayanotoxins I and II (formerly andromedotoxin, rhodotoxin, and acetylandromedol).[1] As many as 18 grayanotoxins (I – XVIII) have been identified, the greatest number being found in the *Leucothoe* species (fetter bush).[2,3]

Figure 339 *Rhododendron exbury* (Azalea)

Figure 340 *Rhododendron* cultivar

Tannins and other compounds are also present in varying amounts. All parts of the rhododendrons including the flowers and the nectar are toxic, although there may be considerable variation between species and even amongst plants of the same species depending on the growing conditions.

Grayanotoxins act to increase sodium channel permeability of cells by opening the channels to sodium, which enters the cells in exchange for calcium ions, thus rendering the channels slow to close so that the cell remains depolarized.[4,5] Other neurologic mechanisms may also involve a cholinergic response seen clinically as bradycardia and excessive salivation.[6] The cardiac effects can range from bradycardia, sinus arrest, and arrhythmias.

Other members of the Ericaceae that contain grayanotoxins include:

Andromeda polifolia	Andromeda, bog rosemary
Kalmia spp.	Laurels
Ledum spp.	Labrador tea
Leucothoe spp.	Fetter bush, dog laurel
Lyonia spp.	Maleberry
Menziesia spp.	Rusty menziesia
Pieris spp.	Pieris, Japanese pieris

Risk Assessment

Rhododendron species are commonly grown as showy garden plants, and azaleas are frequently sold as flowering potted plants for indoor use. Household pets, while not commonly poisoned, are at risk of poisoning if they eat the leaves and flowers. Livestock are poisoned more often where rhododendrons are accessible to the animals, especially in winter time when the evergreen leaves are an attraction. Honey made by bees feeding on the nectar of rhododendrons has long been known to be toxic to people who eat the "mad-honey".[7]

Clinical Signs

Excessive salivation, increased nasal secretions, vomiting, abdominal pain, bloat, and irregular respirations develop several hours after rhododendron leaves are ingested.[8-10] Projectile vomiting may be noticeable. Hypotension, tachycardia, and respiratory depression may also develop. Weakness, partial blindness, and seizures have been reported in severe intoxications. Neurologic signs may persist for several days before the animal recovers. Weight loss may be notable.

Treatment is primarily directed at relief of the more severe clinical signs. Activated charcoal given orally (2-8 g/kg body weight) is helpful if given shortly after the rhododendron is consumed. Atropine is useful in countering the cardiovascular effects

References

1. Kakisawa H, Kozima T, Yanai M, Nakanishi K: Stereochemistry of grayanotoxins. Tetrahedron 21: 3091-3104, 1965.
2. Sakikabara J, Shirai N, Kaiya T, Nakata H: Grayanotoxin-XVIII and its grayanoside B, a new A-Nor-B-Homo-Ent-Kaurine and its glucoside from *Leucothoe grayana*. Phytochemistry 18: 135-137, 1979.
3. Sakikabara J, Shirai N, Kaiya T: Diterpene glycosides from *Pieris japonica*. Phytochemistry 20: 1744-1745, 1981.
4. Narahashi T, Seyama I: Mechanism of nerve membrane depolarization caused by grayanotoxin I. J Physiol 242: 471-487, 1974.
5. Seyama I, Narahashi T: Modulation of sodium channels of squid nerve membranes by grayanotoxin I. J Pharmacol Exp Ther 219: 614-624, 1981.
6. Onat F, Yegan BC, Lawrence R, Oktay A, Oktay S: Site of action of grayanotoxins in mad honey in rats. J Appl Toxicol 11: 199-201, 1991.
7. Krochmal C: Poison honeys. Am Bee J 134: 549-550, 1994.
8. Casteel SW, Wagstaff JD: *Rhododendron macrophyllum* poisoning in a group of goats and sheep. Vet Hum Toxicol 31: 176-1978, 1989.
9. Puschner B et al: Azalea (*Rhododendron* spp.) toxicosis in a group of Nubian goats. Proc 42nd Annual Meeting Am Assoc Vet Lab Diagn. 1999.
10. Frape D, Ward A: Suspected rhododendron poisoning in a dog. Vet Rec 132: 515-516, 1993.

Ricinus communis

Family: Euphorbiaceae

Common Names
Castor bean, castor oil plant, higuerilla, palma Christi

Plant Description
Originating in northeastern Africa and southwestern Asia, *Ricinus* is a monotypic genus that has become widely distributed in most tropical and mild temperate areas of the world. It is grown as a crop plant for its oil, and numerous cultivars have been developed for use as a fast growing ornamental. It is a weed in tropical areas.

Figure 341 *Ricinus communis*

Ricinus communis is a perennial except in temperate areas where it is an annual. It is an erect, branching, fast growing herb that attains heights of 15-20 feet (5-8m) in the tropics. Stems are hollow, hairless, turning red with maturity. Leaves are simple, large, alternate, long petioled, palmate with 5-11 lobes, hairless, glossy

Figure 342 *Ricinus* hybrid spiny seed capsule

Figure 343 Castor beans

green, in some cultivars turning red. Inflorescences are terminal panicles with the staminate (male) flowers on top and the pistillate (female) flowers below. Flowers have 3-5 fused sepals, no petals and many stamens. The fruits are spiny capsules with 3 characteristically mottled seeds. The immature fruits are bright red in some cultivars, turning brown when mature (Figures 341, 342, and 343).

Toxic Principle and Mechanism of Action

Ricin, a glycoprotein (toxalbumin) or lectin present in the seeds, and ricinine, a piperidine alkaloid found in the leaves and seeds are the principle toxic compounds present in the plant.[1,2] Castor oil, commercially extracted from the seeds, contains 90% ricinoleic acid, an unsaturated fatty acid with purgative properties if taken orally. The primary site of action of ricin is on the digestive system epithelium, although the majority of ricin is degraded and passes through the digestive tract with minimal effect. When injected, however, ricin is one of the most toxic biological substances known. Ricin consists of 2 chains of amino acids (A and B chains) linked by a single disulphide bond.[3] The B chain binds to cell receptor sites and facilitates the entry of the A chain into the cell where it enzymatically hydrolyses ribosomal protein thus inhibiting DNA and RNA synthesis.[1] The lectin abrin found in *Abrus precatorius* similarly affects ribosomal protein synthesis. Once bound to cells the ricin inhibits protein synthesis, prevents intestinal absorption, and is directly irritating to the digestive tract causing a hemorrhagic diarrhea.

Ricinine has strong hemagglutinating properties, and may through its action on neuroreceptors be responsible for the seizures and muscular weakness seen in some animals chewing and swallowing castor beans.[4]

Risk Assessment

Castor beans pose the greatest risk to animals and children as the seeds are attractive and are often collected and brought into the domestic environment.[5] Castor bean seed necklaces are commonly acquired by tourists. Intact seeds because of their hard coat will pass through the digestive tract without effect. Seeds that are well chewed and swallowed allow the ricin to exert its toxic effects. Castor bean cake, a product after the castor oil has been extracted, is a source of protein for cattle rations and is toxic unless heat treated. Dogs eating the untreated cake in cattle rations or where it is used as a fertilizer can be poisoned.[6]

Clinical Signs

After a delay of 6 hours or more from the time the seeds were chewed and swallowed, severe diarrhea that may be hemorrhagic is the most common clinical effect. Abdominal pain and straining is common. Vomiting, weakness, dehydration, muscle tremors and sudden collapse may develop in severe cases. Serum liver enzymes are often elevated due to hepatic degeneration. In a series of 98 dogs with castor bean poisoning, the most common signs were diarrhea, vomiting, and depression, with 9% of the cases dying or were euthanized.[7]

If an animal is witnessed eating castor bean seeds, vomiting using apomorphine should be induced as quickly as possible. Treatment of castor bean poisoning should be directed at preventing dehydration and shock. Activated charcoal orally and intravenous fluid and electrolyte therapy should be maintained until the animal's digestive system recovers.

References
1. Lord JM, Roberts LM, Robertus JD: Ricin: structure, mode of action, and some current applications. FASEB J 8: 201-208, 1994.
2. Olsnes S et al. Mecanisms of action of 2 toxic lectins, abrin and ricin. Nature 249: 627-631, 1974.
3. Rutenber E, Robertus JD: Structure of ricin B-chain at 2.5 A resolution. Proteins 10: 260-269, 1991.
4. Ferraz AC et al: Ricinine-elicited seizures. A novel chemical model of convulsive seizures. Pharmacol, Biochem, Behavior 65: 577-583, 2000.
5. Wedin GP, Neal JS. Everson GW, Krenzelok EP: Castor bean poisoning. Am J Emerg Med 4: 259-261, 1986.
6. Soto-Blanco B, Sinhorini IL, Gorniak SL, Schumaher-Henrique B: *Ricinus communis* cake poisoning in a dog. Vet Hum Toxicol 44: 155-156, 2002.
7. Albretsen JC, Gwaltney-Brant SM, Khan SA:Evaluation of castor bean toxicosis in dogs: 98 cases. J Am Anim Hops Assoc 36: 229-233, 2000.

Family: Fabaceae

Figure 344 *Robinia* 'Idaho'

Figure 345 *Robinia* 'Idaho' flowers

Common Names

Black locust, false acacia

Plant Description

Native to North America, *Robinia* is a genus of about 20 species of deciduous, perennial, shrubs and trees attaining heights of 80 feet (24m). The plants spread by suckering and by seed. Trunks are erect, bark gray and deeply grooved, young branches having a pair of spines at each node. Leaves once pinnately compound, 7-20 oval leaflets, turning yellow in the fall. Inflorescences are pendulous racemes from leaf axils. The showy pea-like flowers have 5 fused sepals, 5 petals with banner, and are white in color and fragrant. The fruits are straight leguminous pods containing up to 10 brown kidney-shaped seeds.

Robinia neomexicana, and *R. viscosa* have pink flowers, and have been used in making showy pink flowered hybrids such as "Idaho locust" *(R. pseudoacacia x R. viscose)* (Figures 344-346).

Toxic Principle and Mechanism of Action

Various glycoprotein lectins (toxalbumins) such as robin, and the glucoside robitin have been isolated from the bark and seeds.[1] The specific mechanism of action of the lectins has not been defined but is presumed similar to the lectins ricin and abrin found in castor bean *(Ricinus communis)* and *Abrus precatorius*. Inhibition of protein synthesis and gastrointestinal irritation appear to be the main effects of the toxins

All animal species are susceptible to poisoning including humans, horses, dogs, cats, cattle, poultry, and budgerigars.[1-4]

Risk Assessment

Black locust and its hybrids are often planted for their showy flowers and their drought tolerance. Poisoning of household pets is unlikely, but the trees are a potential risk for livestock and especially horses that chew on the tree's bark out of boredom.

Clinical Signs
Animals eating the bark or seeds of *Robinia* species exhibit gastrointestinal irritation and cardiovascular abnormalities. Vomiting, colic, diarrhea, dehydration, weakness, mydriasis, signs of shock, and cardiac dysrhythmias are commonly seen.[4] Fatalities are uncommon, especially if the animal is treated symptomatically for dehydration

Figure 346 *Robinia pseudoacacia* flowers

and shock. Activated charcoal orally may be helpful in reducing the effects of the lectins, and intravenous fluids and electrolytes should be administered until the animal's gastrointestinal tract has recovered.

References
1. Tazaki K, Yoshida K: The bark lectin of *Robinia pseudoacacia*: purification and partial characterization. Plant Cell Physiol 33: 125-129, 1992.
2. Shropshire CM, Stauber E, Arai M: Evaluation of selected plants for acute toxicosis in budgerigars. J Am Vet Med Assoc 200: 936-939, 1992.
3. Hui A, Marraffa JM, Stork CM: A rare ingestion of the Black Locust tree. J Toxicol - Clinical Toxicology. 42: 93-95, 2004.
4. Burrows GE, Tyrl RJ: Toxic Plants of North America. Iowa State University Press, Ames. pp 602-605, 2001.

Sambucus

Family: Caprifoliaceae

Common Name
Common elderberry, American elder – *Sambucus canadensis*
Red elderberry, elder – *S. racemosa*
Blue elderberry – *S. coerulea.*

Plant Description
A genus of about 25 species of woody shrubs and small trees, *Sambucus* are native to most temperate areas of the world. Erect, branching, deciduous plants with soft woody stems, and pinnate leaves with 5-11 lanceolate leaflets and a

Figure 347 *Sambucus canadensis*

terminal leaflet. Inflorescences are terminal cymes with numerous white or cream flowers, each with 3-5 lobes and prominent stigmas. Fruits are drupes, with 3-5 seeds and either blue-black, red or yellow in color (Figures 347 and 348).

Figure 348 *Sambucus canadensis* berries

Toxic Principle and Mechanism of Action

A variety of toxic substances have been identified in *Sambucus* species including triterpenoids, resins, lectins, and cyanogenic glycosides.[1-3] Rare cases of poisoning have occurred in people possibly due to the irritant triterpenoids.[4] However, cooking and fermenting the ripe berries destroys the toxicity of the compounds.

Risk Assessment

Elderberries are commonly cultivated for their foliage, spring flowers, and colorful edible fruits. Animal poisoning is very unlikely, except possibly in livestock that have access to the plants and other forage is unavailable. Some species of *Sambucus* are quite invasive and may become a problem in pastures.

Clinical Signs

In livestock that have eaten large quantities of elderberry leaves, signs of cyanide poisoning may develop a few hours later. Cherry-red blood, severe respiratory difficulty, and death are typical signs of cyanide poisoning (See Prunus species) Digestive problems including vomiting, colic, and diarrhea are not due to cyanide and are more likely due to other compounds in the plants. These digestive symptoms are usually self limiting, and only require supportive treatment as necessary.

References

1. Inoue T, Sako K: Triterpenoids of *Sambucus nigra* and S. *Canadensis*. Phytochem 14: 1871-1872, 1975.
2. Mach L et al: Purification and partial characterizationof a novel lectin from elder *(Sambucus nigra)* fruit. Biochem J. 278: 667-671, 1991.
3. Seigler DS: The naturally occurring cyanogenic glycosides. In Progress in Phytochemistry vol 4. Reinhold L, Harborne JB, Swain T eds. Pergamon Press, Oxford. 83-120, 1977.
4. Burrows GE, Tyrl RJ: Toxic Plants of North America. Iowa State University Press, Ames. 320-326, 2001.

Sanguinaria canadensis

Family: Papaveraceae

Common Names
Blood root, red puccoon

Plant Description
A genus with one species, *Sanguinaria canadensis* is a springtime perennial native to Eastern North America. The leaves arise from a thick, branching, rhizomatous root that contains a red sap that gives the plant its common name. Leaves are kidney or heart shaped, bluish-green, with 5-7 lobes with crenate edges. The white to pink flowers have 2 sepals and 6-12 unequal, oblong petals with numerous stamens (Figures 349-351). The fruit is a capsule that dehisces from the base.

Figure 349 *Sanguinaria canadensis* leaves

Toxic Principle and Mechanism of Action
Sanguinarine and other protoberberine and benzophenanthridine alkaloids are present in the plant and root.[1] These alkaloids are found in other members of the Papaveraceae including the *Papaver* spp. (poppies), *Argemone* spp. (Mexican or prickly poppy), *Chelidonium majus*

Figure 350 *Sanguinaria canadensis* flowers

(celandine), *Eschscholzia* spp. (California poppy), *Glaucium* spp. (horned poppy), and *Stylophorum* spp. (celandine poppy).[1] Controversy exists as to the toxicity of *Sanguinaria* as it has been used as a medicinal compound.[2] The plant has antibacterial, antitumor, and cytotoxic effects. Because of its antibacterial properties, sanguinarine has been incorporated in some commercial tooth paste products.

Risk Assessment
There is minimal risk of poisoning of pets or children from blood root unless they gain access to the roots and are attracted by the red sap. No cases of animal poisoning have been recently documented

Clinical Signs
Depending on the quantity of the plant or root eaten, clinical signs may include vomiting, abdominal pain, weakness, thirst, and collapse. Depending upon the severity of signs, supportive treatment may be necessary.

References
1. Preininger V: Chemotaxonomy of Papaveraceae and Fumariaceae. In The Alkaloids vol 29, Brossi A ed. Academic Press, Orlando, Florida. pp 1-89, 1986.
2. Duke JA: Handbook of Medicinal Herbs. 6th Ed. CRC Press, Boca Raton, Florida. 424-425, 1988.

Figure 351 *Sanguinaria canadensis* 'grandi flora'

Sapindus
Family: Sapindaceae

Figure 352 *Sapindus saponaria*

Common Names
Soapberry, false dogwood, jaboncillo

Plant Description
A genus made up of 13 species of evergreen or deciduous climbers, shrubs and small trees native to Europe and North America. The bark of the trees is gray-brown and coarsely furrowed. Leaves are pinnately compound, with 7-18 lanceolate leaflets. Inflorescences are large panicles produced at the ends of branches. The flowers are small with 4-5 sepals and petals, and greenish-white in color. The berries are amber in color, translucent, containing large dark brown seeds (Figure 352).

Toxic Principle and Mechanism of Action
A variety of triterpenoid saponins are present in the fruits, and are similar to those found in various holly species (*Hedera* spp.), English ivy *(Hedera helix)*, and common plants such as baby's breath (*Gypsophila* spp.), corn cockle (*Agrostemma githago*), cow cockle *(Vaccaria hispanica)*, poke weed *(Phytolacca americana)*, and bouncing bet *(Saponaria officinalis)*.[1] Saponins (soap-like compounds) are widely distributed in many plant genera. The irritant properties of the saponins can cause gastroenteritis, although ruminants seem to be able to eat the berries without problem.[2]

Risk Assessment
Poisoning from *Sapindus* species is unlikely unless quantities of the berries are consumed. Soapberry is a commonly cultivated tree and the berries persisting on trees in winter are a potential source of poisoning.

Clinical Signs
Vomiting and diarrhea can be anticipated, and in severe cases, symptomatic treatment may be necessary.

References
1. Connolly JD, Hill RA: Dictionary of Terpenoids. Vol 2, Di- and higher terpenoids. Chapman & Hall, London. 1277-1279, 1998.
2. Burrows GE, Tyrl RJ: Toxic Plants of North America. Iowa State University Press, Ames. 1089-1091, 2001.

Saponaria
Family: Caryophyllaceae

Common Names
Bouncing bet, saponaria, soapwort
(S. officinalis)

Plant Description
A genus of about 20 annual and perennial plants native to Europe, Asia, and Africa, *Saponaria* are erect, herbaceous plants forming colonies from a rhizomatous root system. Leaves are petiolate or sessile, opposite, ovate to lanceolate, with 3 parallel veins. Inflorescences are heads of terminal or axillary cymes. Flowers are showy, with 5 sepals in a fused tube, 5 and occasionally 10 petals in white pink or red-purple. Fruits are ovoid, dehiscent capsules with many brown seeds (Figure 353).

Figure 353 *Saponaria officinalis*

Toxic Principle and Mechanism of Action
The seeds especially have many saponins that are gastrointestinal irritants. The saponin glycosides are similar to those found in other plant species such as *Vicaria* spp. (cow cockle, spring cockle), *Agrostemma* species (corn cockle, corn campion), and *Drymaria* species (drymary, inkweed, alfombrillo).[1]

Risk Assessment
Although a common garden plant, poisoning is very unlikely in dogs and cats. The seeds pose the greatest risk, and can be a problem to birds.

Clinical Signs
Vomiting and diarrhea with reddening of the mucous membranes are the most likely signs to be encountered. Livestock eating large quantities plants high in saponins become anorexic, salivate excessively, bloat, and develop diarrhea. Treatment is usually symptomatic.

References
1. Kingsbury JM: Poisonous Plants of the United States and Canada. Prentice-Hall Inc. Englewood Cliffs New Jersey. 1964.

Synonyms: Brassaia, Dizygotheca, Heptapleurum
Family: Araliaceae

Figure 354 *Schefflera actinophylla* in florescence

Figure 355 *Schefflera* hybrid

Common Names
Umbrella tree, octopus tree, Queensland umbrella tree, Australian ivy palm, Hawaiian elf schefflera

Plant Description
A large genus of over 700 species of small trees, shrubs and climbers, *Schefflera* species originate from tropical and subtropical regions of the world. Commonly grown as ornamental garden plants and pocket houseplants, they are prized for their attractive foliage. The alternate, palmate leaves consist of uniform 5-16 leaflets arranged like a cart wheel at the ends of long petioles. The glossy green leaflets vary considerably in size, depending on the species, and in some of the varieties the leaves are variegated. The inflorescences consist of branching radiating spikes, among which the numerous small flowers are arranged, (Figures 354 and 355).

Toxic Principle and Mechanism of Action
Information on the toxic principles present in *Schefflera* species is limited. Compounds similar to those found in *Hedera* species have been found in *Schefflera* and include terpenoids and saponins.[1] Oxalate crystals in the leaves and stems may also be present and contribute to the irritant effects experienced when the plant is chewed.

Risk Assessment
Schefflera are commonly grown as house plants, and as such pose a risk to household pets that chew on the stems and leaves.[2] Poisoning of household pets is rarely reported, and the *Schefflera* species are generally considered one of the safest houseplants.

Clinical Signs
The irritant compounds in the plant will induce vomiting and diarrhea if the leaves and fruits are chewed and swallowed.

Treatment

Unless vomiting and diarrhea are excessive, treatment is seldom necessary. The plant should be removed or made inaccessible to the pets that are eating the plant.

References

1. Burrows GE, Tyrl RJ: (eds) Toxic Plants of North America. Iowa State Press, Ames. pp 120-124, 2001.
2. Stowe CM, Fangmann G: Schefflera toxicosis in a dog. J Am Vet Med Assoc 167: 74, 1975.

Schinus

Family: Anacardiaceae

Common Names

California or Peruvian pepper tree, mastic tree *(Schinus molle)*
Brazilian pepper tree, Florida holly, Christmas berry *(S. terebinthifolius)*. This tree is a noxious weed in Florida.

Plant Description

Consisting of 30 species native to Central America, *Schinus* are invasive, evergreen, branching shrubs or trees that have a resinous sap and a pungent odor to the leaves when crushed. Leaves are alternate, pinnately compound, with 5-15 narrow linear to ovate leaflets. Male and female flowers are produced on separate trees. Small 4-5 yellow or white petalled flowers are produced in terminal or axillary panicles (Figure 356). Female plants produce numerous pea-sized red to purple colored drupes, each containing one seed (Figure 357).

Figure 356 *Schinus molle* flowers

Toxic Principle and Mechanism of Action

All parts of the plants including the berries and especially the sap contain various monoterpenes, triterpenes and phenol/catechol compounds similar to those found in poison ivy *(Toxicodendron)*.

Figure 357 *Schinus terebinthifolius*

On contact with the skin the monoterpene hydroperoxides (alpha-Phellandrene) produce an immediate reaction.[2] There may also be a delayed dermal allergic reaction due to the phenolic and catechol compounds (cardanol).[2] Ingestion of the seeds or plant material will cause intense gastrointestinal irritation. Severe colic in horses, and death in birds has been attributed to the ingestion of *S. terebinthifolius*.[3]

Ingestion of the fruits cause a burning sensation in the mouth that has given the name pink peppercorns to the seeds. However, *Schinus* fruits are not the source of black or white pepper that is commonly used as a spice for cooking. Black pepper is derived from *Piper nigrens*, of the Piperaceae family.

Risk Assessment
Greatest risk is probably to people, especially those allergic to poison ivy, who handle the trees when pruning them. *Schinus terebinthifolius* is a very common and invasive species in Florida and other tropical areas, and was grown initially for its showy red fruits. The trees should be removed from animal enclosures and the seeds should not be fed to birds.[4]

Clinical Signs
Contact with the sap of the plant may result in immediate and/or a delayed allergic dermatitis with intense itching, swelling, and reddening of the skin. If ingested, immediate mouth irritation leading to salivation, vomiting, and diarrhea may occur.
Depending on the amount ingested, symptoms are variable in intensity and may last several days.[3-5]

Treatment should involve washing the skin with a mild soap to remove plant sap, drying agents, and steroids in severe cases. If gastrointestinal signs are present, the use of activated charcoal, and saline cathartics may help in reducing the intestinal irritation.

References
1. Lloyd HA, Jaouni TM, Evans SL, Morton JF: Terpenes of *Schinus terebinthifolius*. Phytochemistry 16: 1301-1302, 1977.
2. Stahl E, Keller K, Blinn C: Cardanol, a skin irritant in pink pepper. Planta Med 48: 5-9, 1983.
3. Morton JF: Brazilian pepper – its impact on people, animals, and the environment. Econ Bot 32: 353-359, 1978.
4. Morton JF: Plants Poisonous to People in Florida. 2nd ed. JF Morton ed. Southeastern Printing, Stuart, Florida. 125-126, 1982.
5. Burrows GE, Tyrl RJ: Toxic Plants of North America. Iowa State University Press, Ames. 34-36, 2001.

Family: Liliaceae

Common Names

Scilla, squill, Commonly cultivated species include autumn squill or starry hyacinth *(S. autumalis)*, Hyacinth scilla *(S. hyacinthoides)*, Cuban lily or Peruvian hyacinth *(S. peruviana)*, bluebell *(S. siberica)*.[1] Other genera closely related to *Scilla*, and at times considered the same genus by some taxonomists, are *Hyacinthoides* (English bluebells) and *Hyacinthus* (hyacinth).

Figure 358 *Scilla siberica*

Plant Description

A genus of some 90 species of bulb forming plants, *Scilla* are native to Europe, Africa and South America. The strap shaped leaves are produced from the bulbs in the spring and are followed by the terminal racemes of blue or white flowers. The individual flowers are small, star-shaped, 6 petalled and clustered terminally on the stems (Figures 358 and 359).

Figure 359 *Scilla peruviana*

Toxic Principle and Mechanism of Action

The bulbs and new growth contain toxic bufadienolides similar to digitalis and similar to scillaren and scilliroside found in Squill or sea onion (*Urginea maritima* L.).[2] Scilliroside is a potent rodenticide and is used as such.

Risk Assessment

Scillas are commonly grown as outdoor plants or as potted indoor plants for their showy blue or white flowers. Poisoning of domestic pets has not been reported to date, but some species are toxic to sheep in areas where the plants are indigenous.[3] The bulbs have the greatest potential for poisoning of people who mistake the bulbs for onions, and pets that have access to the bulbs.

Clinical Signs

Vomiting, salivation, diarrhea, and abdominal pain have been reported in people who ate the bulbs. The cardiac glycosides in the plant can produce severe cardiac dysrhythmias and bradycardia.[4] Hyperkalemia may also develop.

Activated charcoal administered orally with saline cathartics can reduce absorbtion of the toxins. Severe cardiac dysrhythmias and bradycardia will require atropine and other antiarrhythmic drugs. Digitalis specific antibodies may also be effective in severe cases.[5]

References

1. Spoerke DG, Smolinske SC: Toxicity of House Plants. CRC Press. Boca Raton, Florida. pp 198-199, 1980.
2. Stary F: Hamlyn Color Guides Poisonous Plants. Hamlyn, New York., pp 198-199, 1983.
3. Kellerman TS, Coetzer JAW, Naude TW: Plant Poisonings and Mycotoxicoses of Livestock in Southern Africa. Oxford University Press, Capetown. pp 95-96, 1990.
4. Lampe KF, McCann MA. AMA Handbook of Poisonous and Injurious Plants. Am Med Assoc, Chicago, Illinois pp 152-153, 1985.
5. Smith TW et al. Treatment of life-threatening digitalis intoxication with digoxin-specific Fab antibody fragments. Experience in 26 cases. New Engl J Med 307: 1357-1362, 1982.

Senecio

(Packera species)
Family: Asteraceae

Figure 360 *Senecio jacobaea* (tansy ragwort)

Common Names

Cineraria	*Senecio cruentus*
	(Pericallis)
Dusty miller	*S. cineraria*
	(S. bicolor)
String of beads	*S. rowleyanus*
Natal ivy, wax vine	*S. macroglossus*

Many species of *Senecio* are now considered in the genus *Packera.* Although most of the species are wildflowers, some are noxious weeds and are a perennial source of poisoning to animals that eat them. A few species have become popular house and garden plants because of their showy flowers or succulent leaves.

Plant Description

This large and varied genus of over 3000 species is worldwide in distribution. *Senecio* species are either annual, biennial, or perennial forbes, shrubs, climbers, or succulents. Leaves are alternate, varying considerably in shape, being entire or serrated, lobed or pinnately dissected. Flowers are numerous, produced terminally, daisy-like, usually yellow, but can also be white, blue, orange, or red in color (Figures 360-362). Seeds are cylindrical achenes with a hairy pappus that aids in wind distribution.

Toxic Principle and Mechanism of Action

Not all species of *Senecio* are poisonous, but those that are contain toxic pyrrolizidine alkaloids (PA).[1,2] These alkaloids are converted by the liver into toxic pyrroles that inhibit cellular protein synthesis and cell mitosis.[3-5] Hepatocyte necrosis, degeneration, and liver fibrosis with biliary hyperplasia characterize the toxic effects of the pyrrolizidine alkaloids. There is considerable variation in the toxicity of the alkaloids depending upon the species. Horses and cattle are susceptible to PA poisoning, while sheep and goats have rumen microflora that readily transform the PA into non toxic metabolites.[6]

Figure 361 *Senecio cruentus* (Pericallis)

Pyrrolizidine alkaloids are also abundant in the genera Boraginaceae, and Fabaceae. Some of the common species in these genera that contain toxic PAs include *Amsinckia* (Fiddle neck), *Crotolaria* (rattle box), *Echium* (vipers bugloss, blue weed), *Cynoglossum officinale* (hounds tongue), *Symphytum* species (comfrey), and *Heliotropium* (heliotrope).[3,7]

Risk Assessment

The *Senecio* species that are most frequently grown as garden or house plants do not generally contain significant quantities of the toxic pyrrolizidine alkaloids. Furthermore, it is unlikely

Figure 362 *Senecio rowleyanus*

household pets would ingest sufficient quantities of the plant to induce liver toxicity. Herbal products containing comfrey should be used very cautiously because comfrey does contain toxic PA.

Clinical Signs

Animals that consume PA containing plants over a period of time develop signs related to liver failure. Horses and cattle are generally the most severely affected, while sheep and goats are quite resistant to PA toxicity. Weight loss, icterus, diarrhea, photosensitization, and neurologic signs related to hepatic encephalopathy are typical of liver failure. Serum liver enzymes are generally elevated significantly.

Confirmation of PA toxicity can be made by a liver a biopsy showing the triad of histologic changes characteristic of PA poisoning, namely liver megalocytosis, fibrosis, and biliary hyperplasia. Treatment of animals with PA poisoning is generally limited to placing the animal in a barn out of the sun to relieve the photosensitization, providing a high quality, low protein diet, and removing all sources of the PA from the animal's food. The prognosis is generally very poor as once clinical signs of liver failure from PA poisoning occur, the degree of liver damage is severe and irreversible.

References

1. El-Shazly A: Pyrrolizidine alkaloid profiles of some Senecio species from Egypt. J Biosciences 57: 429-433, 2002.
2. Habib AAM: Senecionine, seneciphylline, jacobine and otosenine from *Senecio cineraria*. Planta Medica, 26, 279-282, 1974
3. Cheeke PR: Toxicity and metabolism of pyrrolizidine alkaloids .J Anim Sci. 66: 2343-50, 1988.
4. Mattocks AR: Chemistry and toxicology of pyrrolizidine alkaloids. Academic Press, Orlando Florida 1986.
5. Stegelmeier BL: Pyrrolizidine alkaloids. Clinical Veterinary Toxicology. Ed. Plumlee KH Mosby St Louis, Missouri 2003, 370-377.
6. Craig AM, Blythe LL, Lassen ED, Slizeski ML: Resistance of sheep to p-pyrrolizidine alkaloids. Isr J Vet Med 1986, 42: 376-384.
7. Bull LB, Culvenor CCJ, Dick AT: Frontiers of Biology vol 9, The Pyrrolizidine Alkaloids. North-Holland, Amsterdam 1968, 234-248.

Sesbania

Family: Fabaceae

Figure 363 *Sesbania drummondii*

Common Names

Formerly considered a separate genus, most taxonomists now consider *Daubentonia* species in the genus *Sesbania*. The most common species with the potential for poisoning include:
S. drummondii – coffee or poison bean, Drummond's rattlebush, false poinciana. *S. punicea* – purple sesban or rattlebush, false poinciana

Plant Description

A genus of about 50 species of evergreen, perennial shrubs and small trees, *Sesbania* are native to many tropical and subtropical areas of the world. Leaves are pinnately compound, with 20-70 leaflets, and no terminal leaflet. Inflorescences are axillary racemes of showy, pea-like flowers with 5 fused sepals and 5 petals, yellow, red, purple, white, or combinations thereof (Figures

363 and 364). Fruits are leguminous round or flat pods with numerous seeds.

Toxic Principle and Mechanism of Action

The green and ripe seeds are the most toxic and contain a variety of saponins and cytotoxic compounds including sesbanimide which appears in only the toxic species of *Sesbania*.[1] Some species of Sesbania are used as cattle food, while the seeds of the toxic species have caused poisoning in people, chickens, cattle, and sheep.[2,3] The lethal dose of *S. punicea* seeds is less than 0.1% body weight.[1]

Figure 364 *Sesbania punicea* flowers
(Photographer: Larry Allain USGS)

Risk Assessment

Grown for the attractive foliage and flowers, some *Sesbania* have the potential to cause poisoning in people and animals that eat the seeds.

Clinical Signs

In livestock, signs of anorexia, abdominal pain, rumenal stasis, and weakness develop a day after the plant seeds are eaten. Severe diarrhea then develops causing dehydration. Blood urea nitrogen and hepatic enzymes are frequently increased. Gastrointestinal irritation and kidney tubular degeneration may be observed at post mortem examination.

Children eating the seeds of *S. punicea* develop severe gastrointestinal irritation resulting in diarrhea.[3]

References

1. Powell RG, Plattner RD, Suffness M. Occurrence of sesbanimide in seeds of toxic *Sesbania* species. Weed Sci 38: 148-152, 1990.
2. Flory W, Herbert CD: Determination of the oral toxicity of *Sesbania drummondii* seeds in chickens. Am J Vet Res 45: 955-958, 1945.
3. Morton JF. Plants Poisonous to People in Florida. 2nd ed. JF Morton ed. Southeastern Printing, Stuart, Florida. pp 75-76, 1982.
4. Terblanche M, de Klerk WA, Smit JD, Adelaar TF: A toxicologic study of the plant *Sesbania punicea* Benth. J S Afr Vet Med Assoc 37: 191-197, 1966.

Family: Solanaceae

Figure 365 *Solandra grandiflora*

Figure 366 *Solandra maxima*

Common Names

Chalice vine, cup of gold, trumpet plant, milkcup, Hawaiian lily

Plant Description

A genus of eight species native to tropical America, *Solandra* species are vigorous woody, scrambling plants that are grown for the large fragrant trumpet-shaped flowers. Leaves are shiny, evergreen, leathery, ovate to obovate with entire margins. Flowers are large to 25 cm in length, showy, and fragrant at night. Depending on the species, the flowers are yellow, white, or purpelish in color. The petal lobes are reflexed and fused at their base to form the cup or chalice. Each petal has a purple center-stripe (Figures 365 and 366). The fruit is a conical berry.

Toxic Principle and Mechanism of Action

The principal toxins in the *Solandra* species are the tropane alkaloids scopolamine and hyoscyamine (atropine).[1] The primary effect of the tropane alkaloids is on the autonomic nervous system where they antagonize the action of acetylcholine at the muscarinic cholinergic receptors. Affected animals have dilated pupils, decreased digestive tract motility, and decreased salivation. Respiratory and heart rates are elevated. Hallucinogenic effects can also be expected from the central nervous system effects of the alkaloids.

Risk Assessment

Chalice vines are common in tropical gardens and consequently, the flowers and leaves are accessible to pets. Poisoning has not been reported in domestic animals, although its hallucinogenic properties have led people to consume the plant.[2]

Clinical Signs

Signs of poisoning are due to the effects of tropane alkaloids and generally consist of dilation of the pupils, restlessness, increased heart rate, dyspnea, dry mouth, and intestinal stasis leading to constipation. In severe cases, abnormal behavior due to hallucinations, convulsions, and seizures may occur.

In severe cases, physostigmine, an acetylcholinesterase inhibitor, can be used to reverse the effects of the tropane alkaloids. Usually supportive treatment is all that is necessary.

References
1. Ghani A: The sites of synthesis and secondary transformation of hyoscyamine in *Solandra grandiflora*. Phytochemistry 25: 617-619, 1986.
2. Morton JF: Plants Poisonous to People in Florida and other warm areas. Southeast Printing, Stuart, Florida pp 19-21, 1982.

Solanum pseudocapsicum
Family: Solanaceae

Figure 367 *Solanum pseudocapsicum*

Common Names
Jerusalem cherry, Christmas cherry

Plant Description
Native to the Mediterranean area, this small shrub with lanceolate, v-shaped leaves and prominent mid vein, has white star-like flowers followed by showy round berries that range from yellow to red in color. In cold climates, the plant is an annual (Figure 367).

Toxic Principle and Mechanism of Action
A large number of the *Solanum* species are poisonous to animals and humans because of a variety of glycoalkaloids (solanine) present in all parts of the plant but especially in the unripe fruits.[1] Depending on the species of *Solanum*, even the ripe fruits are poisonous. The glycoalkaloids have an irritant effect on the gastrointestinal system, and depending on the quantity consumed, can cause dilated pupils, tachycardia, and central nervous system depression. In animals the glycoalkaloids predominantly cause digestive symptoms.[2,3]

Figure 368 *Solanum dulcamara* (Bittersweet)

Other commonly cultivated ornamental plants containing glycoalkaloids with potential for causing poisoning include:
Lycium barbarum – matrimony vine, Christmas berry, Chinese boxthorn
Solanum seaforthianum – Potato creeper, Brazilain nightshade, St Vincent lilac
Solanum wendlandii – Potato vine, paradise flower (Costa Rica)

Solandra maxima – Chalice vine, cup of gold, Hawaiian lily
Solanum dulcamara – Bittersweet, woody nightshade (Figure 368)

Risk Assessment

Jerusalem cherries are common as potted houseplants, especially around the holiday season, because of their attractive red berries. The berries are attractive to young children who may pick and eat them.[4,5] Household pets are potentially susceptible to the toxic berries when the plants are accessible to them. Similarly, pet owners should be aware that uncooked, sprouted or green potatoes *(Solanum tuberosum)* also contain glycoalkaloids that are toxic if eaten by pets or children.

Clinical Signs

Colic, vomiting, and diarrhea are the most prevalent signs of poisoning. In large doses the glycoalkaloids can cause depression, rapid heart rate, and respiratory difficulty. Dilated pupils, muscle tremors, and incoordination may also develop in severe poisoning.

Treatment is directed at supportive therapy for vomiting and diarrhea. In severe cases physostigmine may be indicated to reverse the anticholinergic effects of the solanine-type glycoalkaloids

References

1. Burrows GE, Tyrl RJ: Toxic Plants of North America. Iowa State University Press, Ames 1127-1137, 2001.
2. Barr AC: Household and Garden plants. Peterson ME & Talcott, PA eds. Small animal toxicology. Saunders, Philadelphia 263-320, 2001.
3. Evans RJ: Toxic hazards to cats. Vet Annual, 28 , 251-260, 1988.
4. Lawrence RA: Poisonous plants: When they are a threat to children. Pediat Rev 18: 162-168, 1997.
5. Dyer S: Plant exposures: wilderness medicine. Emergency Medicine Clinics North America 22: 299-313, 2004.

Solenostemon

Family: Lamiaceae

Common Name
Coleus, variegated coleus

Plant Description
A genus of some 60 species of shrubby
perennials, *Solenostemon* are native to
tropical Africa and Asia. The stems are
square (4-angled), and the leaves opposite,
toothed, and variegated in colors ranging
from yellow, red, pink, to purple (Figures
369 and 370). Inflorescences are terminal
racemes or panicles, the flowers in whorls.
Individual flowers are small, blue to purple
in color with an elongated lower lip. Fruits
are nutlets enclosed in a papery calyx.

Figure 369 *Solenostemon scutellarioides* cultivars

**Toxic Principle and Mechanism
of Action**
A volatile oil, coleon O, has been isolated
from the leaves and has allergenic
properties.[1,2] Coleus is listed as a
potentially toxic plant by the National
Animal Poison Control Center capable of
causing vomiting, diarrhea, depression,
anorexia, and in severe cases,
hemorrhagic vomiting and diarrhea.[3]

Figure 370 *Solenostemon scutellarioides*

Risk Assessment
Coleus are common tropical plants and are frequently planted as summer annuals
in temperate areas for their colorful variegated leaves. They are also popular
potted house plants. As such, they pose a potential hazard for household pets, but
no known cases of poisoning in animals have been reported. Some people are
prone to developing a contact dermatitis when handling the plants frequently.[2]

Clinical Signs
Vomiting, diarrhea, depression, anorexia, and in severe cases, hemorrhagic
vomiting and diarrhea are the main clinical signs that can be anticipated.[3] When
necessary, symptomatic treatment should be provided.

References
1. Dooms-Goossens A et al: Airborne contact dermatitis to Coleus. Contact
Dermatitis 17: 109, 1987.
2. Spoerke DG, Smolinske SC: Toxicity of House Plants. CRC Press. Boca Raton,
Florida pp 105-107, 1990.
3. Household Plant Reference. ASPCA National Animal Poison Control Center.
424 E. 92nd St., New York, NY 10128

Family: Fabaceae

Figure 371 *Sophora secundiflora*

Figure 372 *Sophora secundiflora* flowers

Figure 373 *Sophora secundiflora* pods

Common Names
Mescal bean, Texas mountain laurel,
Coral bean *(Sophora secundiflora)*
Japanese pagoda tree *(S. japonica)*
Kowhai *(S. tetraptera)*
Silky sophora, white sophora
(S. nuttalliana)

Plant Description
The genus *Sophora* comprises some 70
species found mainly in the northern
hemisphere. Occurring as forbs, shrubs,
or trees, these perennials can be deciduous
or evergreen. Leaves are typically
compound, with 5-60 leaflets, a terminal
leaflet being present. Inflorescences are
terminal racemes or panicles. Sepals are
fused with short lobes and the five petals
are pea-like, with the banner ovate or
obovate, erect or reflexed, the wing petals
oblique-oblong, keels rounded or
accuminate, enclosing the 10 stamens.
Flower colors include white, yellow, and
blue-purple (Figures 371 and 372). The
seeds are produced in legume ponds that
are distinctively constricted between the
seeds, leathery or woody, and either
indehiscent or dehiscent (Figure 373).
The few or many seeds are often yellow or
red in color, waxy, hard-coated, and
oblong or kidney shaped (Figure 374).

Toxic Principle and Mechanism of Action
Sophora species contain a wide variety of
biologically active quinolizidine alkaloids
that are generally grouped according to
their similarity to the alkaloids cytisine,
sparteine, and matrine.[1] Cytisine, being
one of the most toxic alkaloids, has
nicotinic effects on the ganglia, initially
stimulating and then blocking
conduction.[2,3] The alkaloids cause ataxia,
excessive salivation, increased heart rate, and seizures. Cardiac arrhythmias may

develop when large quantities of the alkaloids are consumed. Some of the alkaloids present in *Sophora* species have hallucinogenic properties similar to that of mescaline.[4] The bark of some species *(S. japonica)* contain toxic lectins similar to those found in the bark of the black locust *(Robinia pseudoacacia)*.

Figure 374 *Sophora secundiflora* beans

Risk Assessment
Some species of Sophora are commonly grown as garden plants, and are rarely a problem to household pets. Cattle, sheep, and goats that browse on the plants are more commonly intoxicated. However, some species of Sophora produce large quantities of seed *(S. secundiflora)* and the attractive red waxy beans may be an attraction to pets and children who may ingest them (Figure 374).

Clinical Signs
The nature and severity of clinical signs expected in *Sophora* poisoning depends on the species and the amount of plant ingested. Cattle poisoned by *Sophora secundiflora* develop neurologic signs including stiffness of the hindlegs, muscle tremors, and an inability to rise.[5] Similar neurologic signs of muscle rigidity, exercise intolerance, and collapse on being forced to exercise can be anticipated in dogs and other animals consuming the seeds.[6]

References
1. Burrows GE, Tyrl RJ: Toxic Plants of North America. Iowa State University Press, Ames. pp 612-617, 2001.
2. Barlow RB, McCLeod LJ: Some studies on cytisine and its methylated derivatives. Br J Pharmacol 35: 161-174, 1969.
3. Izaddoost M. Harris BG. Gracy RW: Structure and toxicity of alkaloids and amino acids of *Sophora secundiflora*. J Pharmaceutical Sci. 65: 352-4, 1976.
4. Bourn WN, Keller WJ, Bonfiglio JF: Comparisons of mescaline being alkaloids with mescaline 9-THC and other psychotogens. Life Science 25: 1043-1054, 1979.
5. Sperry OE, Dollahite JW, Hoffman GO, Camp BJ: Texas Poisonous Plants. Texas Ag Exp StnSer B-1028, 1977.
6. Knauer KW. Reagor JC. Bailey EM. Carriker L: Mescalbean *(Sophora secundiflora)* toxicity in a dog. Vet Human Toxicol. 37: 237-9, 1995.

Spathiphyllum
Family: Araceae

Figure 375 *Spathiphyllum* species

Figure 376 *Spathiphyllum* 'Mauna Loa'

Common Name
Peace lily

Plant Description
Originating in tropical America and Malaysia, the 36 species of *Spathiphyllum* are evergreen, rhizomatous perennials that have become universally popular as house and garden plants. The plants are favored for their lush, dark green oval leaves that stand erect or arched, and for the beautiful white, cream, or green flowers carried on long stems above the foliage. The inflorescence consists of a spathe, the true minute flowers are densely arranged on the central spadix, that is surrounded by a large white, cream-colored, or all green spathe (Figures 375 and 376).

Toxic Principle and Mechanism of Action
Like other members of the Araceae, *Spathiphyllum* species contain oxalate crystals within the stems and leaves.[1,2] (See Dieffenbachia spp.)

Risk Assessment
Peace lilies are commonly grown as house plants, and as such pose a risk to household pets that chew on the stems and leaves.

Clinical Signs
Animals that chew on or consume leaves of *Spathiphyllum* species can experience the irritant effects of the oxalates causing excessive salvation and vomiting.

Treatment
Unless salvation and vomiting are excessive, treatment is seldom necessary. The plant should be removed or made inaccessible to the pets that are eating the plant.

References
1. Franceschi VR, Horner HT: Calcium oxalate crystals in plants. Bot Rev 46: 361-427, 1980.
2. Genua JM, Hillson CJ: The occurrence, type, and location of calcium oxalate crystals in the leaves of 14 species of Araceae. Ann Bot 1985, 56: 351-361, 1985.

Sprekelia formossisima

Family: Liliaceae (Amaryllidaceae)

Common Names
Maltese cross lily, Aztec lily, Jacobean lily.

Plant Description
Consisting of a single species indigenous to
Guatemala and Mexico, *Sprekelia* species
grow from a bulb, producing basal narrow,
strap-like leaves, and a single showy red
(rarely white) flower on a 6 inch tall stem.
With 3 bright red sepals and 3 petals, the
bottom 3 forming an open tube, while the
upper 3 curve upwards and outwards
(Figure 377).

Figure 377 *Sprekelia formossisima*

Toxic Principle and Mechanism of Action
Several phenanthridine alkaloids have been identified in the leaves, stems, and
bulbs of *Sprekelia* species.[1] Phenanthridine alkaloids are present in many of the
Liliaceae, most notably in the *Narcissus* group (see *Narcissus*). The alkaloids have
emetic, hypotensive, and respiratory depressant effects, and cause excessive
salivation, abdominal pain, and diarrhea. As many as 15 other phenanthridine
alkaloids have been isolated from other genera of the Amaryllis family, including
species of *Clivia, Galanthus, Haemanthus, Hippeastrum, Hymenocallis, Leucojum,
Narcissus, and Zephranthes.*[1]

Risk Assessment
This attractive garden plant, and occasionally potted houseplant has not been
reported to cause poisoning in animals, but it has the potential to do so.

Clinical Signs
Vomiting, excessive salivation, abdominal pain, diarrhea, and difficulty in
breathing are associated with the phenanthridine alkaloids present in the lily
family. If large quantities of the leaves and bulb are consumed, depression, ataxia,
seizures, and hypotension may develop. Poisoning is rarely fatal, and can generally
be treated symptomatically.

References
1.Martin SF: The Amaryllidaceae alkaloids. In The Alkaloids: Chemistry and
Physiology. vol 30, Brossi A (ed) Academic Press, San Diego, Calif 251-376, 1987.

Symphoricarpos

Family: Caprifoliaceae

Figure 378 *Symphoricarpos albus*

Common Names
Snowberry, waxberry
Commonly cultivated species include
S. albus (snowberry), *S. orbiculatus* (coral berry, buckbush), *S. occidentalis* (wolfberry), *S. oreophilus* (mountain snowberry)

Plant Description
A genus of about 17 species of deciduous shrubs native to North and Central America and China, *Symphoricarpos* are grown for their showy flowers and berries. Branching shrubs forming colonies from sprouting roots, with exfoliating bark, and simple, opposite leaves. The clusters of small flowers are scented, nectar-rich, bell or funnel-shaped, and produced in leaf axils and terminally on branches. Flowers may be white or pink in color. Fruits are berry-like, ovoid, drupes that turn white or purple-red with 2 seeds (Figure 378).

Toxic Principle and Mechanism of Action
The leaves and berries have a reputation of being mildly toxic, but the toxin responsible for the irritant effects of the plant has not been defined. The alkaloid chelidonine has been isolated from the leaves.[1]

Risk Assessment
Generally considered of low toxicity, snow berries have been reported to cause gastrointestinal irritation in children.[2,3] The berries that persist on the shrub after the leaves have fallen are attractive to children and possibly pets.

Clinical Signs
Vomiting and diarrhea are the most common signs associated with snowberries. Signs are usually mild and self limiting. In severe cases, fluid therapy may be necessary to prevent dehydration.

References
1. Szaufer M, Kowalewski Z, Phillipson JD: Chelidonine from *Symphoricarpos albus*. Phytochem 17: 1446-1447, 1978.
2. Lewis WH: Snowberry *(Symphoricarpos)* poisoning in children. J Am Med Assoc 242: 2663, 1979.
3. Lampe KF, McCann MA. AMA Handbook of Poisonous and Injurious Plants. Am Med Assoc, Chicago, Illinois pp 152-153, 1985.

Symphytum

Family: Boraginaceae

Common Names
Common comfrey *(S. officinale)*, Russian comfrey *(S. x uplandicum*, a hybrid of *A. asperum* x *A. officinale)*, rough comfrey *(S. asperum)*, and tuberous comfrey *(S. tuberosum)* are the most common species encountered in North America.

Plant Description
Comprising some 25 species, *Symphytum* species are native to Europe and the Mediterranean area with several species now commonly grown or escaped in North America. Rapidly growing and invasive perennials growing from taproots or rhizomes, erect, branching, up to 200cm in height and generally covered with stiff hairs. Stems may have wings. Leaves are alternate with the uppermost leaves being crowded opposite, petiolate, and lanceolate. Flowers are produced terminally or axillary, on short, branching cymes, each terminating in 5 triangular lobes. Colors vary from blue, pink, cream, to white (Figures 379 and 380). Fruits are 4 ridged nutlets.

Figure 379 *Symphytum officinale*

Figure 380 *Symphytum officinale* flowers

Toxic Principle and Mechanism of Action
A variety of hepatotoxic pyrrolizidine alkaloids (PA) have been identified in the leaves and roots of Russian comfrey *(Symphytum x uplandicum)* and common comfrey *(S. officinale)*.[1-3] Especially in combination, the PA cause chronic hepatotoxicity in animals and humans.[1,4,5] The alkaloids are also carcinogenic and therefore are hazardous to people or animals that are fed medicinal supplements containing comfrey.[6,7]

The PA are converted by the liver into toxic pyrroles that inhibit cellular protein synthesis and cell mitosis.[8,9] Hepatocyte necrosis, degeneration and liver fibrosis with biliary hyperplasia characterize the toxic effects of the pyrrolizidine alkaloids. The alkaloids are passed through the milk of lactating animals and therefore pose a potential risk to suckling animals.[10]

Risk Assessment
There is relatively little risk of comfrey causing poisoning in animals unless they are fed over a period of time supplements containing comfrey. Since the plants can become invasive, care should be taken to see that comfrey does not get into hay meadows, or is fed as garden clippings to corralled livestock.

Clinical Signs

Animals that consume PA containing plants over a period of time develop signs of liver failure. Horses and cattle are generally the most severely affected, while sheep and goats are quite resistant to PA toxicity. Weight loss, icterus, diarrhea, photosensitization, and neurologic signs related to hepatic encephalopathy are typical of liver failure. Serum liver enzymes are generally elevated significantly. Similarly, people who chronically consume comfrey teas and other herbal preparations containing comfrey may develop serious liver veno-occlusive disease.

Confirmation of PA toxicity can be made by a liver a biopsy that shows the triad of histological changes characteristic of PA poisoning, namely liver megalocytosis, fibrosis, and biliary hyperplasia.[11]

Treatment of animals with PA poisoning is generally limited to placing the animal in a barn out of the sun to relieve the photosensitization, providing a high quality, low protein diet, and removing all sources of the PA from the animal's food. The prognosis is generally very poor since once clinical signs of liver failure from PA poisoning occur, the degree of liver damage is severe and irreversible.

References

1. Culvenor CC, et al: Structure and toxicity of the alkaloids of Russian comfrey (*symphytum x uplandicum* Nyman), a medicinal herb and item of human diet. Experientia 36: 377-379, 1980.
2. Couet CE, Crews C, Hanley AB: Analysis, separation, and bioassay of pyrrolizidine alkaloids from comfrey *(Symphytum officinale)*. Natural Toxins 4: 163-167, 1996.
3. Betz JM, Eppley RM, Taylor WC, Andrzejewski D: Determination of pyrrolizidine alkaloids in commercial comfrey products (*Symphytum* sp.). J Pharmaceutical Sci 83: 649-653, 1994.
4. Ridker PN, McDermont WV: Hepatotoxicity due to comfrey herb tea. Am J Med 87: 701 1989.
5. Winship KA: Toxicity of comfrey. Adv Drug Reactions. Toxicol Rev 10: 47-59, 1991.
6. Stickel F, Seitz HK: The efficacy and safety of comfrey. Public Health Nutrition 3(4A): 501-508, 2000.
7. Hirono I, Mori H, Haga M: Carcinogenic activity of *Symphytum officinale.* J National Cancer Instit 61: 865-868, 1978.
8. Cheeke PR: Toxicity and metabolism of pyrrolizidine alkaloids. J Anim Sci. 66: 2343-50, 1988.
9. Mattocks AR: Chemistry and toxicology of pyrrolizidine alkaloids. Academic Press, Orlando Florida, 1986.
10. Panter KE, James LF: Natural plant toxicants in milk: a review. J Anim Sci 68: 892-904, 1990.
11. Stegelmeier BL: Pyrrolizidine alkaloids. Clinical Veterinary Toxicology. Plumlee KH. Ed. Mosby St Louis, Missouri 370-377, 2003.

Syngonium

(*Nephthytis* spp.)
Family: Araceae

Common Names
Arrow head vine, goose foot

Plant Description
A genus of approximately 33 species, *Syngonium* are native to tropical America. The evergreen epiphytic or terrestrial climbing vines have attractive variegated, lobed leaves produced on long stems. The plants contain a milky sap. Flowers are rarely produced in cultivation, and vary in color from white, green, orange, to red (Figure 381). Similar in appearance to the *Syngonium* species are the *Scindapsus* species (Figure 382).

Figure 381 *Syngonium* 'Painted Arrow Illusion'

Toxic Principle and Mechanism of Action
Like other members of the Araceae, *Syngonium* contain oxalate crystals within the stems and leaves.[1,2] (See *Dieffenbachia* spp.)

Risk Assessment
Arrowhead vines are commonly grown as house plants, and as such pose a risk to household pets that chew on the stems and leaves.

Figure 382 *Scindapsus* spp.

Clinical Signs
Animals that chew on or consume leaves of *Syngonium* species can experience the irritant effects of the oxalates causing excessive salvation and vomiting.

Treatment
Unless salvation and vomiting are excessive, treatment is seldom necessary. The plant should be removed or made inaccessible to the pets that are eating the plant.

References
1. Franceschi VR, Horner HT: Calcium oxalate crystals in plants. Bot Rev 46: 361-427, 1980.
2. Genua JM, Hillson CJ: The occurrence, type, and location of calcium oxalate crystals in the leaves of 14 species of Araceae. Ann Bot 56: 351-361, 1985.

Family: Taxaceae

Figure 383 *Taxus baccata*

Common Name
Yew
North American indigenous species of
yew include:

Taxus brevifolia	Western or Pacific yew
T. canadensis	Canada yew, American yew, ground hemlock
T. floridana	Florida yew

Introduced species of yew include:

T. baccata	English yew
T. chinensis	Chinese yew
T. cuspidate	Japanese yew

Numerous hybrids have been developed
from indigenous and introduced species.

Plant Description
A genus of 5-10 species of evergreen
shrubs or trees which are native to most
northern temperate zones and extending
into Central America and the
Philippines. *Taxus* species are branching
shrubs or trees ranging from 2-25m in
height. The bark is scaly and reddish
brown. Leaves are dark green above, paler
underneath, arranged spirally. Pollen

Figure 384 *Taxus baccata* berries

cones are produced on the male plants while the female plants produce fleshy, red
to orange, single seeded, cup-shaped arils (Figures 383 and 384).

Toxic Principle and Mechanism of Action
Yews contain a variety of toxins including cyanogenic glycosides, taxane
derivitives, ephedrine, irritant oil, and molting hormone.[1,2] Only the nitrogenous
taxane alkaloids known as taxines are associated with the cardiotoxic effects of the
Taxus species. Taxine B is more toxic than taxine A, and the proportions of the
taxines varies depending upon the species of *Taxus*.[3,4] Taxine B causes decreased
cardiac contractility and marked slowing of atrial and ventricular rates. This effect
is mediated through taxine B's inhibition of sodium, potassium, and calcium
channels in cardiac muscle.[5] All parts of the plant are toxic except for the ripe red-
orange aril. The seed itself is toxic. The plant also appears to be more toxic in
winter.[6] *Taxus canadensis* has the lowest concentrations of taxines and probably
has little potential for toxicity. All domestic animals including birds are

susceptible to poisoning by taxine B.[7-11] Deer and moose appear to be far less susceptible to poisoning, apparently because their rumen microflora are able to detoxify taxines.[12]

Taxol has antineoplastic properties, and has been used as a chemotherapeutic agent in treating certain types of cancer. Its effects are attributed to its inhibition of cell mitosis through the promotion of microtubule formation and stability.[12] Taxol is found in *T. brevifolia* but is also present in other species. It has also been synthesized.[1,14]

Risk Assessment

Yews are one of the most toxic plants that are commonly planted for their dense green evergreen foliage that is ideal for hedges and pruning into ornamental shapes. In many instances, yew poisoning of animals occurs when animals are accidentally fed yew leaves after the plants are pruned. At other times horses or livestock gain access to the yew in gardens. As little as 200g of yew is lethal to a 500kg horse.[14] Humans are susceptible to yew poisoning with approximately 100g of the leaves being lethal.[13] An ounce of the leaves is lethal to dogs. No species of *Taxus* should be planted in or around animal enclosures, and it is important not to dispose of yew trimmings into animal pens.

Clinical Signs

Horses and cattle that have eaten yew are often found dead with no other signs being noticed prior to death.[16,17] Animals may die suddenly several days after they have consumed yew. Weakness, ataxia, muscle tremors, nervousness, bradycardia, jugular distension, and difficulty in breathing may all be present in an animal affected by yew poisoning.[6,7] Dogs eating yew leaves vomit, develop muscle tremors, seizures, panting, dilated pupils, and marked increase in heart rate.[9] Stressing the animal can precipitate sudden collapse and death. There are generally no gross post mortem lesions.

Cases of human poisoning are rare and result in vomiting, diarrhea, delirium, and convulsions.[18] Death occurs as a result of bradycardia and complete heart standstill.

Treatment must be initiated shortly after the yew is consumed to ensure success. Activated charcoal orally, and a rumenotomy to remove recently eaten yew from the rumen may be lifesaving. Gastric lavage may be attempted to try and remove yew leaves from the stomach. Atropine may be helpful in counteracting the bradycardia if administered early.

Intravenous lidocaine hydrochloride has been used in a case of human poisoning to control ventricular fibrillation. Seizuring dogs may need to be given pentobarbital or given diazepan to control seizures.[9,10]

A diagnosis of yew poisoning in animals is generally made by finding evidence of the plant in the stomach or rumen of the animal, absence of other pathology, and the detection of the taxine alkaloids in the stomach contents using liquid chromatography and mass spectrometry.[4]

References

1. Khan NUD, Parveen N: The constituents of the genus *Taxus*. J Sci Indus Res 46: 512-516, 1997.
2. Jenniskens LHD, van Rosendaal ELM, van Beek TA, Wiegerinck PHG, Scheeren HW: Identifaction of six taxine alkaloids from *Taxus baccata* needles. J Nat Prod 59: 117-123, 1996.
3. Adeline MT, Wang XP, Poupat C, Ahond A, Potier P: Evaluation of taxoids from *Taxus* sp. Crude extracts by high performance liquid chromatography. J Liq Chromatogr Relat Technol 20: 3135-3145, 1997.
4. Kite GC, Lawrence TJ, Dauncey EA: Detecting *Taxus* poisoning in horses using liquid chromatography/mass spectrometry. Vet Human Toxicol 42: 151-154, 2000.
5. Tekol Y, Kameyama M: Electrophysiological studies of the mechanism of action of the yew toxin taxine on the heart. Arzneim-Forsch 37: 428-431,1987.
6. Alden CL, Fosnaugh CJ, Smith JB, Mohan R: Japanese yew poisoning of large domestic animals in the Midwest. J Am Vet Med Assoc 170: 314-316, 1977.
7. Panter KE, Molyneux RJ, Smart RA, Mitchell L, Hansen S: English yew poisoning in 43 cattle. J Am Vet Med Assoc 202: 1476-1477, 1993.
8. Cope RB, Camp C, Lohr CV. Fatal yew (*Taxus* sp) poisoning in Willamette valley,Oregon, horses. Vet Human toxicol 46: 279-281, 2004.
9. Evans KL, Cook JR: Japanese yew poisoning in a dog. J Am Anim Hosp Assoc 27: 300-302, 1991.
10. Cope RB: The Dangers of yew ingestion. Vet Med 9: 646-650, 2005.
11. Lowe JE, Hintz HF, Schryver HF, Kingsbury JM: *Taxus cispidata* (Japanese yew) poisoning in horses. Cornell Vet 60: 36-39, 1968.
12. Weaver JD, Brown DL: Incubation of European yew *(Taxus baccata)* with white-tailed deer *(Odocoileus virginianus)* rumen fluid reduces taxine A concentrations. Vet Human Toxicol 46: 300-302, 2004.
13. Schiff PB, Fant J, Horwitz SB: Promotion of microtubule assembly in vitro by taxol. Nature 277: 665-667, 1979.
14. Witherup KM, Look SA, Stasko MW, Ghiorzi TJ, Muschik GM: *Taxus* spp. needles contain certain amounts of taxol comparable to the bark of *Taxus brevifolia*: analysis and isolation. J Nat Prod 53: 1249-1255, 1990.
15. Wilson CR, Sauer J, Hooser SB: Taxines: a review of the mechanism and toxicity of yew (*Taxusi* spp) alkaloids. Toxicon 39: 151-154, 2001.
16. Rook J. Japanese yew toxicity. Vet Med 89: 950-951, 1994.
17. Helman RG, Fenton K, Edwards WC, Panciera RJ, Burrows GE: Sudden death in calves due to Taxus ingestion. Agri-Practice 17: 16-18, 1996.
18. Krenzelok EP, Jacobsen TD, Aronis J; Is the yew really poisonous to you? J Toxicol Clinical Toxicol 36: 219-223, 1998.

Thalictrum

Family: Ranunculaceae

Common Name
Meadow rue

Plant Description
Some 300 species of *Thalictrum* occur throughout the northern temperate zone. Branching perennials, growing from woody rhizomes or tuberous roots, *Thalictrum* have pinnate, compound leaves, and colorful flowers in terminal panicles or racemes (Figures 385 and 386).

Toxic Principle and Mechanism of Action
Thalictrum species contain a large number of alkaloids and glycosides in addition to the irritant glycoside ranunculin that is converted to protoanemonin when the plant tissues are chewed and macerated.[1] Protoanemonin levels amongst the species vary and appear to be low.[2] Protoanemonin is a vesicant, and it is polymerized to the non toxic anemonin.

Figure 385 *Thalictrum fendlerii*

Figure 386 *Thalictrum dasycarpum* flowers

Risk Assessment
Meadow rues are popular garden plants, but are not a problem to household pets as the bitter, irritant effects of the plants make the plant unpalatable. The large numbers of alkaloids present in the plant provide the potential for poisoning.

Clinical Signs
Excessive salivation, vomiting, and diarrhea can be anticipated if buttercups are eaten.
Treatment if necessary would be symptomatic.

References
1. Hill R, Van Heyningen R: Ranunculin: the precursor of the vesicant substance I Buttercup. Biochem J 49: 332-335, 1951.
2. Bonora A, Dall'Olio G, Bruni A: Separation and quantification of protoanemonins in Ranunculaceae by normal and reversed phase HPLC. Planta Med 51: 364-367, 1985.

Theobroma

Family: Sterculiaceae

Figure 387 *Theobroma cacao* flowers and fruit

Figure 388 *Theobroma cacao* pod

Common Name

Cocoa

Plant Description

A genus of about 20 species of evergreen trees from tropical America, *Theobroma cacao* is the best known for its seeds that are the source of chocolate. It is an erect, branching tree up to 9m. in height. Leaves are large, ovate, glossy green, leathery, and up to 25cm in length. Small creamy-yellow flowers are produced on the woody trunk and branches. Fruits are large ribbed pods up to 30cm in length that ripen to a reddish brown, and contain numerous seeds (Figures 387 and 388).

Toxic Principle and Mechanism of Action

Cocoa seeds contain the methylxanthine alkaloids theobromine and caffeine. These alkaloids are rapidly absorbed from the digestive tract and act to block the adenosine receptor which is critical to cyclic AMP and other cell functions.[1,2] Consequently, an increase in heart rate, bronchodilation and in high doses, convulsions can result from these alkaloids. The methylxanthines may also exert their effects by increasing intra cellular calcium and levels of cyclic AMP with profound effects on many cell functions.[1]

Most poisoning results from dogs and cats eating chocolate, foods containing chocolate, or even the hulls from the cocoa beans.[1-5] Dark chocolate can contain up to 134mg/oz of theobromine and 35mg/oz of caffeine in contrast to milk chocolate that may contain up to 58mg/oz of theobromine and 6mg/oz of caffeine.[2] The toxic (LD50) of theobromine in dogs is 250-500mg/kg body weight.[2] In other words a 10kg dog could be poisoned by eating about 18oz of dark chocolate. Cats and even wild carnivores are susceptible to theobromine/caffeine poisoning.[7]

Risk Assessment

Since dogs find chocolate quite palatable and chocolate is commonly available in the household, the potential for poisoning is always present. Cocoa bean hulls have been used as garden mulches and in at least one case dogs became intoxicated by eating the cocoa hulls.[5, 6]

Clinical Signs

Vomiting, diarrhea, increased urination, hyperexcitability, fast heart and respiratory rates are signs of poisoning in dogs. An increased body temperature, muscle tremors, and seizures may also occur. Hematuria, bradycardia, coma, and death may occur in severe cases.[1] Dogs known or suspected of eating chocolate should be evaluated by a veterinarian.

Dogs known to have eaten quantities of chocolate, particularly dark (cooking chocolate) in the recent past should be induced to vomit. Gastric lavage with warm water may help to remove the chocolate from the stomach. Activated charcoal and a saline cathartic is indicated if the chocolate was eaten more than 2 hours prior to the onset of signs.

If the animal is seizuring, diazepam may be effective initially but may become ineffective, and phenobarbital or pentobarbital may be necessary to control the seizures. Cardiac irregularities if pronounced will need appropriate treatment. Intravenous fluid therapy should be used supportively to enhance diuresis.[1]

References

1. Hooser SB, Beaseley VR: In Current Veterinary Therapy IX. Kirk RW ed. WB Saunders, Philadelphia 191-192, 1966.
2. Albretsen JC: Methylxanthines. In Clinical Veterinary Toxicology. Plumlee KH ed. Mosby,St Louis, Missouri 322-325, 2003.
3. Albretsen JC, Gwaltney-Brant SM, Khan SA: Cocoa poisoning in a dog. Vet Rec 109: 563-564, 1981.
4. Stidworthy MF, Bleakley JS, Cheeseman MT, Kelly DF: Chocolate poisoning in dogs. Vet Rec 141: 28, 1997.
5. Drolet R, Arendt TD, Stowe CM: Cacoa bean shell poisoning in a dog. J Am Vet Med Assoc. 185: 902, 1984.
6. Hansen SR et al: Cocoa bean mulch as a course of methylxanthine toxicosis in dogs. Clinical Tox. 41(5), 2003.
7. Jansson DS, Galgan V, Schubert B, Segerstad CH: Theobromine intoxication in a red fox and a European badger in Sweden. J Wildlife Dis 37: 362-365, 2001.

Family: Apocynaceae

Figure 389 *Thevetia thevetiodes*

Figure 390 *Thevetia peruviana*

Common Names

Yellow oleander, be-still tree, lucky nut, tiger apple - *Thevetia peruviana (Cascabela thevetia)*
Giant yellow oleander, yoyote - *T. thevetiodes*

Plant Description

A genus of 8 species of shrubs or small trees native to the tropics of Central America and West Indies, *Thevetia* species are commonly cultivated for their showy yellow flowers. Typically, the plants are perennial, evergreen, branching shrubs and small trees, with simple, linear, alternate, glossy, green leaves and a milky sap. The flowers are produced terminally, and are funnel-shaped, fragrant, showy, with sepals fused basally, and petals ranging from bright yellow to orange or pinkish-yellow. Fruits are green angular drupes that turn red to black when ripe (Figures 389 and 390).

Toxic Principle and Mechanism of Action

All animals, including humans, consuming *Thevetia* species are susceptible to the potent cardiotoxic glycosides Thevitin A and B and thevetoxin that are present in all parts of the plants and especially the fruits and seeds.[1-4] The glycosides act directly on the gastrointestinal tract causing hemorrhagic enteritis, abdominal pain, and diarrhea.[5] The cardiac glycosides act by inhibiting the cellular membrane sodium-potassium (Na^+-K^+ ATPase enzyme system) pump with resulting depletion of intracellular potassium and an increase in serum potassium.[5] A progressive decrease in electrical conductivity through the heart causes irregular heart activity, and eventually complete block of cardiac activity. Toxic doses of the glycosides cause a variety of severe dysrhythmias and conduction disturbances through the myocardium that result in decreased cardiac output, heart block, and death.

Risk Assessment

Thevetia species are common garden plants in tropical and sub-tropical areas, and like oleander *(Nerium oleander)* are one of the more poisonous of plants that people and animals are exposed to. In some parts of the world, the fruits of *Thevetia* are used for suicidal purposes.[6] Extracts from the leaves of the plant are also toxic to fish.[7]

Clinical Signs

Animals consuming Thevetia species may be found dead due to the profound effects of the cardiotoxins. A variety of cardiac dysrhythmias and heart block including first and second degree heart block and ventricular tachycardia may be encountered with cardiac glycoside poisoning. Abdominal pain (colic), and diarrhea are also signs commonly encountered in animals poisoned with cardiac glycosides. If observed early in the course of poisoning, animals will exhibit rapid breathing, cold extremities, and a rapid weak and irregular pulse. The duration of symptoms rarely exceeds 24 hours before death occurs.

In acute cardiac glycoside poisoning, the post mortem findings include hemorrhages, congestion, edema, and cell degeneration of the organs of the thoracic and abdominal cavities. In less acute but fatal poisoning, tissue necrosis resulting from decreased oxygenation may be present in a variety of organs.

Treatment

Gastric lavage or vomiting should be induced in dogs and cats as soon as possible. Activated charcoal orally appears to improve survivability.[8] The cardiac irregularities may be treated using anti-arrhythmic drugs such as potassium chloride, procainamide, lidocaine, dipotassium EDTA, or atropine sulfate.[9] The use of fructose-1,6-diphosphate (FDP) has been shown to effectively reduce serum potassium levels, and irregularities of the heart (dysrhythmias), and will improve cardiac function in dogs experimentally poisoned with oleander and presumably in *Thevetia* poisoning.[10] The mechanism of action of FDP is not known but it apparently restores cell membrane Na^+ and K^+ ATPase function.[10] As hyperkalemia is a common feature of oleander poisoning, the administration of potassium containing fluids should be done very cautiously, and not at all unless serum potassium levels can be monitored closely. Intravenous fluids containing calcium should not be given as calcium augments the effects of the cardiac glycosides. Poisoned animals should be kept as quiet as possible to avoid further stress on the heart.

References

1. Langford SD, Boor PJ: Oleander toxicity: an examination of human and animal toxic exposures. Toxicology. 109: 1-13, 1996.
2. Ansford AJ, Morris H: Oleander poisoning. Toxicon 3: 15-16, 1983.
3. Pawha R, Chatterjee VC. The toxicity of yellow oleander *(Thevetia neriifolia Juss)* seed kernels to rats. Vet Hum Toxicol 32: 561-564, 1990.
4. Oji O, Okafor QE: Toxicological studies on stem bark, leaf and seed kernel of yellow oleander *(Thevetia peruviana).* Phytotherapy Res 14: 133-135, 2000.
5. Radford DJ, Gillies AD, Hinds JA, Duffy P: Naturally occuring cardiac glycosides. Medical J Australia 144: 540-544, 1986.
6. Eddleston M, et al: Epidemic of self-poisoning with seeds of the yellow oleander tree *(Thevetia peruviana)* in northern Sri Lanka. Tropical Medicine & International Health. 4: 266-273, 1999.
7. Sambasivam S, Karpagam G, Chandran R, Khan SA: Toxicity of leaf extract of yellow oleander *Thevetia nerifolia* on Tilapia. J Environmental Biology. 24: 201-204, 2003.

8. de Silva HA Et al: Multiple-dose activated charcoal for treatment of yellow oleander poisoning: a single-blind, randomized, placebo-controlled trial. Lancet. 361: 1935-1938, 2003.
9. Adams HR. Digitalis and vasodilator drugs. In: Veterinary Pharmacology and Therapeutics. 7th ed. Adams HR.(ed) Iowa State University Press, Ames, pp 451-481, 1995.
10. Markov AK, et al: Fructose-1,6-diphosphate in the treatment of oleander toxicity in dogs. Vet Human Toxicol 41: 9-15, 1999.

Toxicodendron Mill.

Family: Anacardiaceae

Figure 391 *Toxicodendron radicans*

Figure 392 *Toxicodendron pubescens* (Poison ivy)

Common Names

Poison ivy *(Toxicodendron radicans)*, eastern poison ivy *(T. pubescens)*, western poison ivy *(T. rydbergii)*, poison oak *(T. diversilobium* or *T. quercifolium)*, and poison sumac, poison ash *(T. vernix)*

Plant Description

There are 6-10 species of *Toxicodendron* native to North America and parts of Asia, 5 species of which are variously named and found in North America. Depending upon the area and growing conditions, there are close similarities between the species making their identification difficult at times. The leaves are glossy, hairless, alternate, always three-parted, and may be entire, toothed, or lobed. The climbing stems produce aerial roots that anchor the plant to the bark of trees on which it is climbing. Species growing in the drier western States tend to be more shrub-like rather than climbers. The flowers are small, whitish, hanging clusters, and the fruits are white and berry-like with longitudinal grooves. The leaves turn a bright red in the Fall (Figures 391, 392, 393, 394, and 395).

Toxic Principle and Mechanism of Action

Poison ivy, poison oak, and poison sumac all contain highly irritating allergenic phenolic compounds collectively called urushiol (oleoresin).[1] The most toxic being the resin 3-n-pentadecylcatechol.[2] All parts of the plant, green or dried contain urushiol. The oily resin is not volatile or soluble in water but is soluble in alcohol. Smoke from burning the plants can contain droplets of the toxin and will affect people who are highly allergic to the toxin. Urushiol coming into contact with human skin, penetrates and binds to cell proteins that then induces an immune response, the severity of which depends on the sensitivity of the individual. The cell-mediated immune response may take 1-3 days to develop in people who have had previous exposure to urushiol. Sensitization to urushiol can be induced by plants other than *Toxicodendron* species that contain the urushiol eg: mango *(Mango indica)*, cashew *(Anacardium occidentale)*, Brazilian pepper tree *(Schinus terebinthifolius)*, and skunkbush sumac *(Rhus trilobata).*[3]

Figure 393 *Toxicodendron diversilobium* (Poison oak)

Animals themselves are rarely if ever affected by the urushiol, but their hair coat may become contaminated with urushiol when they contact the plants, and therefore people who handle the animals may become exposed to the toxin. Goats and sheep have been observed eating poison ivy without apparent problems.

Risk Assessment

Figure 394 *Toxicodendron vernix* (Poison sumac)

Poison ivy and its related species are common indigenous plants that are often present in gardens especially those adjacent to forested areas and open spaces where the plants can become established. Recognition and awareness of poison ivy and its look-alikes is important in order to avoid human exposure. The rule of thumb "leaves of three, leave it be" is a good one, however, there are several harmless plants that have three leaves that mimic poison ivy that are worth differentiating. These include boxelder *(Acer negundo)*, white ash seedlings *(Fraxinus americana)*, Virginia creeper *(Parthenocissus quinquefolia)*, and Boston ivy *(Pathenocissus tricuspidata).*[4]

Clinical Signs
Skin reddening, swelling, and intense itching where ever the skin is in contact with the urushiol. In some individuals fluid filled vesicles form, rupture, and

Figure 395 *Toxicodendron radicans* - fall color

form scabs. Contrary to popular rumor the fluid from the vesicles does not contain urushiol and does not spread the rash. In highly sensitized individuals, systemic signs of vomiting, diarrhea, and abdominal pain may develop. Without therapy the dermatitis may persist for several weeks.

Treatment

As soon as there is known contact with poison ivy, the skin should be liberally washed with water. A mild soap that does not remove all the protective skin oils may be helpful in removing the urushiol. Once urushiol has bound to the cell membranes (about 10 minutes of contact time), it cannot be removed easily.[5] Once dermatitis and itching have developed, soothing topical drying agents such as calamine may help reduce the severity of signs.[6,7] Numerous 'remedies' have been recommended for treating poison ivy dermatitis including such things as applying the juice of jewel weed (*Impatiens* spp.), but none have been proven to be universally effective.[8]

In severe cases, of poison ivy dermatitis, a person should see a physician because the most effective treatment may be the use of the more potent systemic corticosteroids.[5]

References

1. Gross M. Baer H, Fales HM: Urushiols of poisonous *Anacardiaceae*. Phytochemistry 14: 2263-2266, 1975.
2. Baer H: Allergic contact dermatitis from plants. In: Plant and fungal toxins. Vol 1. Keeler RF, Tu AT (Eds) Marcel Dekker Inc. New York 421-442, 1983.
3. Tucker MO, Swan CR: Images in clinical medicine. The mango-poison ivy connection. New England J Med 339: 235, 1988.
4. McGovern TW, LaWarre SR, Brunette C: Is it, or isn't it? Poison ivy look-a-likes. Am J Contact Dermatitis 11: 104-11, 2000.
5. Williford PM, Sheretz EF: Poison ivy dermatitis. Nuances in treatment. Archives Family Medicine 3: 184-188, 1994.
6. Guin JD: Treatment of toxicodendron dermatitis (poison ivy and poison oak). Skin Therapy Letter 6: 3-5, 2001.
7. McGuffey EC: What methods are effective to prevent or treat poison ivy/oak/sumac? American Pharmacy 33: 18, 1993.
8. Long D, Ballentine NH, Marks JG: Treatment of poison ivy/oak allergic contact dermatitis with an extract of jewelweed. Am J Contact Dermatitis 8: 150-153, 1997.

Tulipa

Family: Liliaceae

Common name
Tulip.
Most of today's tulip hybrids were developed from the ancient cultivar *Tulipa gesneriana*.

Plant Description
Native to central and western Asia, the 100 or so *Tulipa* species and their many hybrids are widely grown in most temperate parts of the world. Arising from an onion-like, brown, pointed tipped bulb, the erect stems with 2-4 clasping, basal leaves range in height from 10cm to 70cm depending on the species and cultivar. Leaves are thick, fleshy, oblong to ovate, and covered by a powdery substance that can be rubbed off. Flowers are single or in clusters, showy, consisting of 6 similar petal-like perianth parts, cup or bell-shaped, or star-like in shape. Flowers come in a wide range of colors from black to bronze, to white, yellow, red, or purple (Figures 396 and 397). Fruits are globose or ellipsoid capsules with many flat seeds.

Figure 396 *Tulipa* hybrids

Figure 397 *Tulipa* hybrid

Toxic Principle and Mechanism of Action
All parts of the plant and especially the bulbs contain glycosides tuliposide A and B in the cells which when ruptured are converted to the irritant or allergenic lactones tulipin A and B. A lectin and a glycoprotein have also been identified that may be responsible for the toxicity of tulips.[1,2]
People who handle the bulbs a lot may develop a dermatitis due to the calcium oxalate raphide crystals that penetrate the skin and cause inflammation.[2-4]

Risk Assessment
Tulip bulbs pose the greatest risk to people and animals that may handle or eat the bulbs. The bulbs pose the greatest risk to dogs.
A contact dermatitis or allergy occurs in some individuals handling the tulip bulbs.[2-4]
Poisoning has occurred in people who have mistakenly eaten tulip bulbs for onions.[5] Fatalities in people who have eaten tulip bulbs have not been reported. However, cattle fed quantities of discarded tulip plants and bulbs have been fatally poisoned.[6]

Clinical Signs

Vomiting, increased salivation, increased heart rate, difficulty in breathing , and occasionally diarrhea can result. In cattle that have been fed tulip bulbs, intestinal irritation, excessive salivation, decreased feed digestion, loss of weight, regurgitation of rumen contents, diarrhea, and death may result.[6]

Erythema, alopecia, and pustular lesions may develop in some people who handle the bulbs frequently. A similar contact dermatitis may also occur in some individuals who handle other common bulbs or plants of *Hyacinthus*, and *Allstroemaria* species.[4]

Treatment, when necessary, should include administration of activated charcoal orally, and supportive treatment if diarrhea and vomiting is severe. A physician should be consulted where dermatitis persists.

References

1. Oda Y, Minami K: Isolation and characterization of a lectin from tulip bulbs, *Tulipa gesneriana*. Eur J Biochem 159: 239-246, 1986.
2. Spoerke DG, Smolinske SC: Toxicity of House Plants. CRC Press. Boca Raton, Florida. pp 212-214, 1990.
3. Hausen BM: Airborne contact dermatitis caused by tulip bulbs. J Am Acad Dermatol. 7: 500-504, 1982.
4. Hjorth N, Wilkinson DS; Contact dermatitis. IV. Tulip fingers , hyacinth itch, lily rash. Br J Dermatology 80: 696-698, 1968.
5. Maretic Z, Russel FE, Ladavac J: Tulip bulb poisoning. Period Biol 80: 141-143, 1978.
6. Wolf P, Blanke HJ, Wohlsein P, Kamphues J, Stober M: Animal nutrition for veterinarians--actual cases: tulip bulbs with leaves *(Tulipa gesneriana)* -- an unusual and high risk plant for ruminant feeding. DTW - Deutsche Tierarztliche Wochenschrift 110: 302-305, 2003.

Vinca

Family: Apocynaceae

Common name
Common periwinkle, old maid, running myrtle – *Vinca minor*
Blue or greater periwinkle – *Vinca major*

Plant Description
A genus of 8 species native to Europe and North Africa, *Vinca* species are perennials with trailing stems containing a milky sap. Leaves are simple, opposite, 2-3 cm long, dark green, glossy surfaced, round-ovate, and pinnately veined. The flowers are showy, produced in alternate leaf axils, and have a short tube opening to 5 flat petals that range from a rosy purple to blue in color. (Figure 398) Fruits are short follicles with numerous seeds.

Figure 398 *Vinca minor*

Closely related to *Vinca* is the genus *Catharanthus* (Madagascar periwinkle). Some taxonomists consider these two plant genera one and the same! Originating in Madagascar, and now cosmopolitan, it is a popular garden plant. It has a compact habit and is a profuse bloomer in white, pink, rosy red, magenta or purple, some with a red center.

Toxic Principle and Mechanism of Action
Numerous alkaloids including alstronine, reserpine, vinblastine, vincristine, and yohimbine are present in all parts of the plant and have hypotensive, digestive and neurotoxic effects if consumed in large doses.[1,2] Some alkaloids such as reserpine reduce blood pressure, while vinblastine and vincristine affect bone marrow cells and have been used effectively for the treatment of leukemias.[3] These alkaloids inhibit mitosis through binding of tubulin and arresting microtubule formation. They also have neurotoxic and teratogenic effects in mice.[4] Cattle and sheep have reportedly developed neurotoxicity from grazing the plant.[5]

Risk Assessment
Vinca or periwinkle is commonly grown as a garden plant for its attractive flowers and as a ground cover. It also does well as a potted plant. Because of its high alkaloid content there is potential for poisoning of household pets that might eat the plant.

Clinical Signs
Animals eating the leaves in quantity develop anorexia, anemia, hypotension, incoordination, muscle tremors, lateral flexion of the neck, and convulsions. Coma and death follow.[5,6]
There is no specific treatment and animals should be given symptomatic supportive treatment including activated charcoal orally and intravenous fluid therapy.

References

1. Taylor WI. The *Vinca* alkaloids. In The Alkaloids. Vol 8. Manske RHF ed. Academic Press, New York 1965, 269-285.
2. van Der Heijden R. Jacobs DI. Snoeijer W. Hallard D. Verpoorte R. The *Catharanthus* alkaloids: pharmacognosy and biotechnology. Current Medicinal Chemistry. 2004,11:607-628.
3. Noble RL. The discovery of the vinca alkaloids – chemotherapeutic agents against cancer. Biochem Cell Biol 1990, 68:1344-1351.
4. Zimmerman JL, Todd GC, Tamura RN. Target organ toxicity and leukemia in Fisher 34 rats given intravenous doses of vinblastine and descacetyl vinblastine. Fund Appl Toxicol 1991, 17:482-493.
5. Selva Raj VB, Ganapathy MS. Studies on the toxicity of *Lochnera pusilla* K Schum. Indian Vet J 1967, 44:871-876.
6. Burrows GE, Tyrl RJ. Toxic Plants of North America. Iowa State University Press, Ames. 2001, 77-78.

Vitis

Family: Vitaceae

Figure 399 *Vitis venifera*

Figure 400 *Vitis venifera*

Common Name
Grape, grape vine.

Plant Description
A genus of about 65 vines or shrubs found in many regions of the Northern Hemisphere, *Vitis* is best known for *Vitis venifera*, the wine grape. Numerous cultivars of this species are the basis for the varietal wines available today. A frost-hardy, woody, deciduous vine with tendrils opposite the leaves, and large 5-7 lobed, to heart shaped leaves. Inflorescences are produced in leaf axils and consist of clusters of small greenish-white, 5 petalled flowers. Fruits are drupes, with or without seeds, that turn blue-black or yellowish-green when ripe (Figures 399 and 400).

Toxic Principle and Mechanism of Action
The toxin present in grapes or raisins responsible for the syndrome of poisoning in dogs has not been determined.[1-4] Tannins in the grapes and raisins, fungal mycotoxins in the grapes such as ochratoxin, and pesticides that may have been sprayed on the grapes have been postulated as a possible cause

for acute renal failure seen in dogs, but no evidence to date has been found to incriminate these compounds.[3,4] The high glucose and fructose content of grapes (12%) and raisins (40%) suggests that dogs may be unusually sensitive to mono-saccharides or compounds that interfere with sugar metabolism, and the syndrome may be due to the high sugar intake affecting kidney function.[5] There is considerable variation in the susceptibility of dogs to grape and/or raisin toxicity. In one series of suspected grape or raisin toxicity in dogs the amount of grapes consumed was 12-31gm/kg body weight.[3] The toxic dose of raisins is 0.1oz/kg body weight.

Risk Assessment
Dogs that have eaten grapes or raisins they have been fed in the household are at greatest risk. Consumption of grapes off of the vines directly by dogs can lead to toxicity. Toxicity does not appear to be related to the variety of grape, or brand of raisin consumed.

Clinical Signs
Vomiting within 24 hours of consuming grapes or raisins is the initial presenting sign in dogs.[3,4] Even though it would appear most or all of the grapes or raisins have been vomited, vomiting continues for several hours and leads to dehydration, weakness, and collapse. This is followed within a short period by signs of acute oliguric or anuric renal failure. Increased blood urea nitrogen (BUN), and creatinine, hyperphospatemia, and hypercalcemia are generally present within 48 hours of ingesting the raisins ir grapes.[5-7] Severe, diffuse, tubular necrosis is often detectable on renal biopsy.

Successful treatment of raisin or grape poisoning requires early recognition and aggressive monitoring and appropriate therapy to maintain renal function. If the grapes or raisins have been consumed in the past 2 hours, vomiting should be induced, and activated charcoal administered orally. Dogs should be put on intravenous fluid therapy, and closely monitored since dogs with a history of ingesting grapes or raisins are often in acute oliguric or anuric renal failure when they are presented for treatment, caution must be taken when administering intravenous fluids as the dogs can quickly develop pulmonary edema. Peritoneal dialysis may be necessary and life saving.[6,7]

References
1. Campbell A, Bates N: Raisin poisoning in dogs. Vet Rec 152: 376, 2003.
2. Penny D, Henderson SM, Brown PJ: Raisin poisoning in a dog. Vet Rec 152: 308, 2003.
3. Gwaltney-Brant S, Holding JK, Donaldson CW, Eubig PA, Khan SA: Renal failure associated with ingestion of grapes or raisins in dogs. J Am Vet Med Assoc 218: 1555-1556, 2001.
4. Mazzaferro EM, et al: Acute renal failure associated with raisin or grape ingestion in 4 dogs. J Vet Emerg Crit Care 14: 203-212, 2004.
5. Singleton VL: More information on grape toxicosis. J Am Vet Med Assoc 219: 434-435, 2001.
6.Eubig PA et al: Acute renal failure in dogs after the ingestion of grapes or raisins: A retrospective evaluation of 43 dogs (1992-2002) J. Vet Intern Med. 19: 663-674, 2005.
7. Porterpan B: Raisins and grapes: potentially lethal treats for dogs Vet Med 5: 346-350, 2005.

Wisteria
Family: Fabaceae

Figure 401 *Wisteria sinensis alba*

Figure 402 *Wisteria sinensis*

Common Name
Wisteria, Chinese wisteria *(W. sinensis)*, Japanese wisteria *(W. floribunda)*, Silky wisteria *(W. brachybotrys)*, American wisteria *(W. frutescens)*

Plant Description
Comprising 10 species native to Asia and North America, *Wisteria* species are vigorous, deciduous, woody stemmed climbers to 10-15m in length. Leaves are pinnate, with 9-15 lanceolate or elliptic leaflets. Inflorescences are many, pendant racemes up to 60cm in length, white or purple, pea-like, perfumed flowers (Figures 401-403). Fruits are leguminous pods with multiple seeds.

Toxic Principle and Mechanism of Action
The seeds and bark contain a toxic glycoprotein lectin which binds to N-acetyl-D-galactosamine and thereby prevents the normal replacement of the mucosal cell layer of the intestinal mucosa.[1] The loss of the mucosal cells results in hemorrhagic gastroenteritis and corresponding clinical signs. Although the flowers are reportedly edible, caution is advisable as all parts of the plants are potentially toxic.[2]

Risk Assessment
Wisterias are popular garden plants for growing over arbors and as espaliers. Once mature they produce striking displays of flowers, followed by bean like pods that contain toxic seeds. The seeds are hazardous to children and pets that might consume them.

Clinical Signs
Gastroenteritis, hematemesis, diarrhea, headache, dizziness, and confusion are reported in people who have ingested 5-10 seeds.[3] Similar clinical signs can be expected in animals that eat the seeds.

Since vomiting is a common effect of *Wisteria* poisoning, the seeds are often regurgitated, and treatment must be directed at fluid replacement where dehydration is present.

References
1. Kurokawa T, Tsuda M, Sugino Y: Purication and characterization of a lectin from *Wisteria floribunda* seeds. J Biol Chem 251: 5686-5693, 1976.
2. Hedrick UP: (Ed) Sturtevant's edible plants of the world. Dover Publications, New York, NY. 1972.
3. Rondeau ES: *Wisteria* toxicity. J Toxicology - Clinical Toxicology 31: 107-112, 1993.

Figure 403 *Wisteria sinensis* flowers

Xanthosoma

Family: Araceae

Common Names
Elephant's ear, cocoyam, malanga, tannia

Plant Description
Comprising a genus of 45 or more tropical American species of tuberous perennials, *Xanthosoma* species are cultivated both as ornamentals and for their edible roots. The leaves are usually soft, large, arrow-shaped with petioles from 1-3 ft in length, attached at the leaf cleft edge, and with downward pointing blades. Inflorescences are frequently absent or inconspicuous and consists of a vertical spadix surrounded by a creamy-white spathe. The tuberous

Figure 404 *Xanthosoma violaceum*

rhizome is elongated white or yellow, with a starchy texture. The tuberous roots of some species are edible when cooked. *Xanthosoma* species are similar in appearance to *Alocasia* and *Colocasia* species, but the leaf petioles in the latter attach to the underside of the leaf midway between the leaf cleft and the center of the leaf (Figure 404).

Toxic Principle and Mechanism of Action
Like other members of the Araceae family, *Xanthosoma* species contain oxalate crystals in the stems and leaves. The calcium oxalate crystals (raphides) are contained in specialized cells referred to as idioblasts.[1,2] Raphides are long needle-like crystals grouped together in these specialized cells, and when the plant tissue

is chewed by an animal, the crystals are extruded into the mouth and mucous membranes of the unfortunate animal. The raphides once embedded in the mucous membranes of the mouth cause an intense irritation and inflammation. Evidence exists to suggest that the oxalate crystals act as a means for introducing other toxic compounds from the plant such as prostaglandins, histamine, and proteolytic enzymes that mediate the inflammatory response.[3]

Risk Assessment
Xanthosoma species are often grown as ornamentals for their showy leaves in tropical gardens and as house plants in temperate areas. Consequently they have the potential for causing poisoning in household pets that might chew the stems and leaves.

Clinical Signs
Dogs and cats that chew repeatedly on the leaves and stems may salivate excessively and vomit as a result of the irritant effects of the calcium oxalate crystals embedded in their oral mucous membranes. The painful swelling in the mouth may prevent the animal from eating for several days. Severe conjunctivitis may result, if plant juices contact the eye.

Treatment
Unless salvation and vomiting are excessive, treatment is seldom necessary. Anti inflammatory therapy may be necessary in cases were stomatitis is severe. The plant should be removed or made inaccessible to the animals that are eating the plant.

References
1. Franceschi VR, Horner HT: Calcium oxalate crystals in plants. Bot Rev 46: 361-427, 1980.
2. Genua JM, Hillson CJ: The occurrence, type, and location of calcium oxalate crystals in the leaves of 14 species of Araceae. Ann Bot 56: 351-361, 1985.
3. Saha BP, Hussain M: A study of the irritating principle of aroids. Indian J Agric Sci 53: 833-836, 1983.

Yucca

Family: Agavaceae

Common Names
Yucca, Spanish bayonet, Spanish dagger

Plant Description
A genus of some 40 species indigenous to the drier regions of North America, yuccas are well adapted to dry conditions, and therefore are popular xeriscape plants. They form rosettes of stiff, sword-like leaves often tipped with a sharp spine. Some species develop a woody, branching, trunk as they mature. Inflorescences consist of tall, showy panicles of pendent, creamy-white, bell-like flowers (Figure 405). Some of the

Figure 405 *Yucca* species

larger species of yucca may take many years before they bloom. Many species of yucca require a specific insect, the yucca moth, to pollinate the flowers. The fruits are fleshy capsules that turn brown when ripe, splitting open to release numerous seeds. Other members of the Agavaceae that are similar include agave (*Agave* spp.) and sacahuiste, bunch grass, or bear grass (*Nolina* spp.).

Toxic Principle and Mechanism of Action
Members of the Agavaceae contain steroidal saponins that cause a crystalloid cholangiohepatopathy. As a result of the precipitation of the calcium saponins in the bile ducts, the liver is unable to excrete phylloerythrin normally, and hepatogenous photosensitivity results. Poisoning is most likely in ruminants that might graze the plants when there is little else available.

Risk Assessment
Yuccas have minimal risk of being poisonous to household pets. Some species, however have terminal spines that can cause mechanical injury.

Clinical Signs
Vomiting and diarrhea could be anticipated in a dog or cat that chewed upon or swallowed yucca leaves. The problem is usually self limiting. Weight loss, and photosensitivity suggestive of liver disease can be expected in ruminants that graze yucca in dry range conditions. Animals with photosensitization should be kept in the shade and removed from the source of the yucca.

References
1. Flaoyen A, Wilkins AL, Sandvik M: Rumenal metabolism in sheep of saponins from *Yucca schidigera*. Vet Res Communications 26: 159-69, 2002.

Family: Zamiaceae

Figure 406 *Cycas revoluta*

Figure 407 *Zamia integrifloia*

Common Names
Sago palm, coontie, chamal, coyolillo, cycad

Plant Description
Comprising about 60 species, *Zamia* is a diverse genus of the cycad family found in Florida, Caribbean area and tropical South America. There is one species that is native to the states of Florida and Georgia, *Zamia integrifolia (Z. floridana)* that is commonly known as coontie, Florida arrow root, Seminole bread or guayiga.
Similar genera are *Ceratozamia, Cycas, Dioon,* and *Macrozamia* (Figure 406).

Zamia integrifolia is a perennial herb growing from an underground, branching stem, with 4-10 erect leaves up to 4 feet (1.5 m) in length, with up to 80 dark green, glossy, stiff, linear leaflets. The male pollen cones are narrowly cylindrical, while the female seed cones are ovoid, both being a rusty brown color. The seeds are bright orange to red in color and are exposed when the fruit is ripe (Figures 407-409).

Toxic Principle and Mechanism of Action
The primary toxins in the *Zamia* species are the glycosides cycasin and macrozamin, found in all parts of the plant but especially in the seeds. The glycosides are hydrolyzed by bacterial enzymes in the digestive tract of animals consuming the plant or its seeds to produce the toxic aglycone methylazoxymethanol (MAM), which alkylates DNA and RNA causing severe hepatic necrosis.[1,2] It also has carcinogenic, mutagenic, and teratogenic properties.[2,3] Similar toxins are found in the fruits of the *Cycas* and *Macrozamia* species.[4]

A second toxin, an amino acid beta-N-methylamino-L-alanine (BMAA), is found predominantly in the seeds and causes neurologic lesions similar to other neurolathyrogens involving neuroreceptors.[5] Prolonged consumption of the *Cycas* seeds or flour made from the seeds is necessary to induce neurologic signs and lesions.

Risk Assessment

Dogs and humans are most likely to be affected after eating the attractive fruits that are toxic. *Zamia* species are commonly grown as garden plants, in tropical areas, and occasionally as house plants in temperate areas.

Clinical Signs

Humans, dogs, cattle, and sheep that eat the seeds of *Zamia* species typically develop signs of poisoning within an hour of consuming the seeds, and exhibited signs of digestive disturbance, liver failure, and in some instances neurological signs.[6-10]

Figure 408 *Zamia* species fruit cone

Dogs typically vomit and may continue doing so for several hours. Excessive salivation and increased thirst are often noticeable. During the next few days anorexia, depression, diarrhea or constipation, and icterus develop. Serum bilirubin, liver enzymes, blood urea nitrogen, and creatinine levels become elevated. The prognosis is guarded to poor once evidence of hepatic necrosis develops. Neurological signs are more common in people and livestock who have consumed the plants or seeds over a

Figure 409 *Zamia floridiana* fruits

prolonged period. Neurologic signs include weakness, loss of proprioception, staggering gait, and incoordination. Treatment should be directed to support the liver and neurological signs.

At post mortem examination, the characteristic findings include an enlarged, congested liver, with a "nutmeg" appearance. Focal and centrilobular necrosis of the liver is seen in acute cases, while fibrosis and biliary hyperplasia is seen in more chronic cases. In cattle showing neurologic signs, bilateral, symmetrical degeneration of the cervical spinal cord is evident.[9]

References

1. Yagi F, Tadera K: Azoxyglycoside contents in seeds of several cycad species and various parts of the Japanese cycad. Agric Biol Chem 51: 1719-1721, 1987.
2. Spatz M, Smith DWE, McDaniel EG, Laquer GL: Role of intestinal microorganisms in determining cycasin toxicity. Proc Soc Exp Biol Med 124: 691-697, 1967.
3. Keeler RF: Known and suspected teratogenic hazards in range plants. Clin Toxicol 5: 529-565, 1972.

4. Mills JN, Lawley MJ, Thomas J: Macrozamia toxicosis in a dog. Austr Vet J 73: 69-72, 1996.
5. Nunn PB, Davis AJ, O'Brien P: Carbamate formation in the neurotoxicity of L-alpha-amino acids. Science 251: 1619, 1991.
6. Albretsen JC. Khan SA. Richardson JA. Cycad palm toxicosis in dogs: 60 cases (1987-1997). J Am Vet Med Assoc 213: 99-101, 1998.
7. Senior DF et al: Cycad poisoning in the dog. J Anim Hosp Assoc 21: 103-109, 1985.
8. Gabbedy BJ, Meyer EP, Dickson J: Zamia palm *(Macrozamia reidlei)* poisoning of sheep. Aust Vet J 51: 303-305, 1975.
9. Reams RY, et al: Cycad *(Zamia puertoriquensis)* toxicosis in a group of dairy heifers in Puerto Rico. J Vet Diagn Invest 5: 488-494, 1993.
10. Hall WT. Cycad (zamia) poisoning in Australia. Austr Vet J 64: 149-151, 1987.

Zantedeschia

Family: Araceae

Figure 410 *Zantedeschia aethiopica*

Common Name
Calla lily, arum lily, pig lily, lirio calla

Plant Description
Consisting of 6 species, *Zantedeschia* are native to East and Southern Africa. They have become a popular garden ornamental or household plants because of their showy flowers and attractive foliage. *Zantedeschia* species are perennials in tropical and subtropical areas, and some species remain green year round provided they are given adequate moisture at all times. Leaves are glossy green, leathery, lanceolate, ovate-cordate, with long petioles that are sheathing at the base. The leaf blade margins are often undulating and in some species, white markings are present on the leaves. The fluorescence is a showy white, yellow, or pink spathe, shaped like a funnel with a central yellow spadix. Hybrids of the species have been developed that have spathes, ranging in color from yellow, red to purple (Figures 410, 411, and 412). The fruits are yellow or red berries.

Toxic Principle and Mechanism of Action
Members of the Araceae family contain oxalate crystals in the stems and leaves.[1] The calcium oxalate crystals (raphides) are contained in specialized cells referred to as idioblasts.[2,3] Raphides are long needle-like crystals bunched together in these specialized cells, and when the plant tissue is chewed by an animal, the crystals are extruded into the mouth and mucous membranes of the unfortunate animal. The

raphides once embedded in the mucous membranes of the mouth cause an intense irritation and inflammation. Evidence exists to suggest that the oxalate crystals act as a means for introducing other toxic compounds from the plant such as prostaglandins, histamine, and proteolytic enzymes that mediate the inflammatory response.[4] (See *Dieffenbachia* species)

Risk Assessment
Zantedeschia species are frequently grown as potted house plants for their showy flowers and foliage and consequently, household pets have access to the plants. Unless a household pet is a particular plant eater, poisoning from these plants is rare. In one report, Zantedeschia was incriminated in the deaths of children in Brazil who had eaten the plant's flowers.[5]

Clinical Signs
Dogs and cats that chew on the leaves and stems of *Zantedeshia* may salivate excessively and vomit as a result of the irritant effects of the calcium oxalate crystals embedded in their oral mucous membranes. The painful swelling in the mouth may prevent the animal from eating for several days. Severe conjunctivitis may result, if plant juices are rubbed in the eye.

Treatment
Unless salvation and vomiting are excessive, treatment is seldom necessary. Anti inflammatory therapy may be necessary in cases where stomatitis is severe. The plant should be removed or made inaccessible to the animals that are eating them.

Figure 411 *Zantedeschia elliottiana*

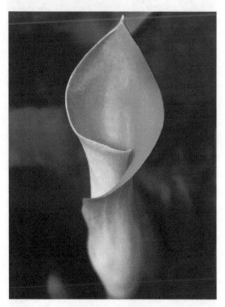

Figure 412 *Zantedeschia* hybrid

References
1. Lin T-J et al: Calcium oxalate is the main toxic component clinical presentations of *Alocasia macrorrhiza* (L) Schott & Endl poisonings. Vet Hum Toxicol 14: 93-95, 1998
2. Franceschi VR, Horner HT: Calcium oxalate crystals in plants. Bot Rev 46: 361-427, 1980.

3. Genua JM, Hillson CJ: The occurrence, type, and location of calcium oxalate crystals in the leaves of 14 species of Araceae. Ann Bot 56: 351-361, 1985.
4. Saha BP, Hussain M: A study of the irritating principle of aroids. Indian J Agric Sci 53: 833-836, 1983.
5. Ladiera M, Andrade SO, Sawaya P: The toxicity of household plants: a twenty five year prospective study. Toxicol Appl Pharmicol 34: 373-373, 1975.

Zephyranthes

Family: Liliaceae

Figure 413 *Zephyranthes grandiflora*

Figure 414 *Zephyranthes atamoasco*

Common Names

Rain lily, atamasco lily, fairy lily, wind flower, zephyr lily, storm lily, Brazos rain lily, prairie lily, evening-star rain lily

The genus *Cooperia* is now considered to be synonymous with *Zephyranthes*. The pampas lily (*Habranthus* spp.) is also a synonym for *Zephyranthes* species.

Plant Description

The genus *Zephyranthes* consists of some 70 species, indigenous to the warm temperature climates of the southeastern United States and the Americas. Growing from onion-like bulbs covered with a papery tunic, the plant has glossy, light green, linear, basal leaves, and produces a single starry white to pink flower per stem. Some species have dark pink and yellow flowers. The flowers tend to close at night (Figures 413 and 414).

Toxic Principle and Mechanism of Action

Although the specific toxins have not been identified, *Zephyranthes* species contain phenanthridine alkaloids including lycorine, galantine, and tazettine.[1] The phenanthridine alkaloids are present in many of the Liliaceae, most notably in the Narcissus group. The bulbs appear to be the most toxic, and in one report, 0.5% of an animal's body weight of the bulbs was found to be lethal for cattle.[2]

A poorly defined photosensitivity has been reported in cattle that have consumed the dry vegetative parts of *Zephyranthes drummondii (Cooperia pedunculata)* especially after the dried leaves are rained upon.[3]

Risk Assessment

Zephyranthes have not been reported as a problem for household pets. However, with the ever increasing introduction of attractive plants into the horticultural market, there is the potential for the toxic bulbs to be accessible to household pets.

Clinical Signs

Vomiting, excessive salivation, abdominal pain, diarrhea, and difficulty in breathing are associated with the phenanthridine alkaloids present in the lily family. If large quantities of the leaves and bulb are consumed, depression, ataxia, seizures, and hypotension may develop. Treatment is necessary should be directed towards relieving diarrhea and dehydration.

References

1. Martin SF: The Amaryllidaceae alkaloids. In The Alkaloids: Chemistry and Physiology. vol 30, Brossi A (ed) Academic Press, San Diego, Calif pp 251-376, 1987.
2. Kingsbury JM: poisonous plants of the United States and Canada. Englewood Cliffs, New Jersey. Prentice-Hall, pp 469, 1964.
3. Rowe LD, Norman JO, Corrier DE et al: photosensitive station of Capitol Southeast Texas: identification of phototoxic activity associated with *Cooperia pedunculata*. Am J Vet Res 48: 1658-1661, 1987.

A

Achene – A dry indehiscent one-seeded fruit, attached to the pericarp at only one place; formed from a single carpel, the seed is distinct from the fruit, as in Asteracea.

Acuminate – Gradually tapering sides finished before arriving at the apex or tip.

Acute – Sharp pointed.

Alternate – With a single leaf or other structure at each node.

Annual – Living one growing season.

Articulate – Jointed; breaking into distinct pieces without tearing at maturity.

B

Berry – Simple, fleshy indehiscent (not splitting open) fruit with one or more seeds (tomato, nightshade).

Biennial – Living two growing seasons.

Bipinnate – Twice pinnately compound.

Blade – The expanded part of a leaf or floral part.

Bract – A small, rudimentary or imperfectly developed leaf.

Bulb – A bud with fleshy bracts or scales, usually subterranean.

C

Calyx – The outer set of sterile, floral leaves called sepals.

Campanulate – Bell-shaped.

Canescent – Becoming gray or grayish.

Capitate – Arranged in a head, as the flowers in compositae.

Capsule – A dry fruit of two or more carpels, usually dehiscent by valves.

Carpel – A portion of the ovary or female portion of the flower.

Catkin – Spike-like inflorescence, unisexual, usually with scaly bracts.

Caulescent – Having a stem.

Cillia – Fine hairs or projections.

Ciliate – Having fine hairs or projections, usually as marginal hairs.

Compound – Composed of several parts or divisions.

Cordate – Heart-shaped.

Corolla – The inner set of sterile, usually colored, floral leaves; the petals considered collectively.

Corymb – A raceme with the lower flower stalks longer than those above, so that all the flowers are at the same level.

Cuneate – Wedge-shaped.

Cuspidate – having a rigid point.

Cyathium – a modified inflorescence comprising a pistillate flower arising from a cup-like involucre with glands on the rim of the cup and staminate flowers on the inner surface as in the *Euphorbia* species.

Cyme – An inflorescence; a convex or flat flower cluster, the central flowers unfolding.

D

Deciduous – Dying back; seasonal shedding of leaves or other structures;falling off.

Decumbent – Lying flat, or being prostrate, but with the tip growing upwards.

Dentate – Toothed, with outwardly projecting teeth.

Denticulate – Finely toothed. Diffuse – loosely spreading.

Dioecious – Only one sex in a plant; with male or female flowers only.

Disk – (disc) – A flattened enlargement of the receptacle of a flower or inflorescence; the head of tubular flowers, as in sunflower.

Dissected – Divided into many segments.

Drupe – A fruit with a fleshy or pulpy outer part and a bone-like inner part; a single seeded fleshy fruit. Can with or without seeds

Druplet – A small drupe, as one section of a blackberry.

E

Elliptic – Oval.

Entire – Without teeth, serrations, or lobes, as in leaf margins.

F

Fascicle – A cluster of leaves or other structures croweded on a short stem.

Fibrous – A mass of adventitious fine roots.

Filiform – Threadlike.

Follicle – A many-seeded dry fruit, derived from a single carpel , and splitting longitudinally down one side.

Frond – Large, compound, much divided leaf as in ferns, cycads or palms

Fruit – The ripened ovary or ovaries with the attached parts fuscous- Dingy brown.

G

Glabrate – Nearly without hairs.

Glabrous – Smooth or hairless.

Glaucous – Covered with bluish or white bloom.

Globose – globular or spherical

Glume – Small dry, membranous bract at the base of a grass spikelet.

H

Hastate – Arrow-shaped with the basal lobes spreading.

Head – A dense inflorescence of sessile or nearly sessile flowers, as in Compositae.

Hirsute – Having rather course, stiff hairs.

I

Indehiscent – Not opening at maturity.

Inflorescence – The arrangement of flowers on the flowering shoot, as a spike, panicle, head, cyme, umbel, raceme.

Involucre – Any leaflike structure protecting the reproducing structure, as in flower heads of *Compositae* and *Euphorbiaceae*.

K

Keel – Projecting, united front petals as in the flowers of *Fabiaceae* (peas).

L

Lanceolate – Flattened, two or three times as long as broad, widest in the middle and tapering to a pointed apex; lance-shaped.

Leaf sheath – The lower part of a leaf, which envelopes the stem, as in grasses.

Leaflet – One of the divisions of a compound leaf.

Legume pod – A dry fruit, splitting by two longitudinal sutures with a row of seeds on the inner side of the central suture; as in family *Fabaceae (Leguminosae)*.

Lenticular – Bean-shaped; shaped like a double convex lens.

Ligule – A membrane at the junction of the leaf sheath and leaf base of many grasses.

Linear – A long and narrow organ with the sides nearly parallel.

Lobed – Divided to about the middle or less.

M

Midrib – The central rib of a leaf or other organ; midvein.

Monoecious – Having seperate male and female flowers on the same plant.

N

Node – The part of a stem where the leaf, leaves, or secondary branches emerge.

Nutlet – A one-seeded portion of a fruit that fragments at maturity.

O

Obcordate – Inversely heart-shaped.

Oblanceolate – Inversely lanceolate.

Oblique – With part not opposite, but slightly uneven.

Oblong – Elliptical, blunt at each end, having nearly parallel sides, two to four times as long as broad.

Obovate – Inversely ovate.

Obtuse – Blunt or rounded.

Ocrea – A thin, sheathing stipule or a united pair of stipules (as in *Polygonaceae*).

Orbicular – Nearly circular in outline.

Ovate – Egg-shaped.

P

Palmate – Diverging like the fingers of a hand.

Panicle – A inflorescence, a branched raceme, with each branch bearing a raceme of flowers, usually of pyramidal form.

Pappus – A ring of fine hairs developed from the calyx, covering the fruit; acting as a parachute for wind-dispersal, as in dandelion.

Pedicel or peduncle – A short stalk.

Peltate – More or less flattened, attached at the center on the underside.

Perennial – Growing many years or seasons.

Perfect – A flower having both stamens and carpels.

Perfoliate – Leaves clasping the stem, forming cups.

Perianth – The calyx and corolla together; a floral envelope.

Pericarp – The body of a fruit developed from the ovary wall and enclosing the seeds.

Persistent – Remaining attached after the growing season.

Petal – One of the modified leaves of the corolla; usually the colorful part of a flower.

Petiole – The unexpanded portion of a leaf; the stalk of a leaf.

Pilose – Having scattered, simple, moderately stiff hairs.

Pinnate – Leaves divided into leaflets or segments aling a common axis; a compound leaf.

Pistillate – Female-flowered, with pistils only.

Prickle – A stiff, sharp-pointed outgrowth from the epidermis, as in Solanum.

Procumbant – Lying on the ground.

Puberulent – With very short hairs; woolly.

Pubescent – Covered with fine, soft hairs.

Punctate – With translucent dots or glands.

R

Raceme – An inflorescence, with the main axis bearing stalked flowers, these opening from the base upward.

Racemose – Like a raceme or in a raceme.

Rachis – The axis of a pinnately compound leaf; the axis of inflorescence; the portion of a fern frond to which the pinnae are attached.

Ray – A marginal flower with a strap-shaped corolla, as in *Compositae*.

Receptacle – The end of the flower stalk, bearing the parts of the flower.

Reniform – Kidney-shaped. Reticulate-netted, as veins in leaves; with a network of fine upstanding ridges, as on the surface of spores.

Rhizome – An elongated underground stem, as in ferns.

Rootstock – An elongated underground stem, usually in higher plants.

Rosette – A cluster of leaves, usually basal, as in dandelion.

S

Sagittate – Arrowhead-shaped.

Scale – A highly modified, dry leaf, usually for protection.

Scape – A leafless or nearly leafless stem, coming from an underground part and bearing a flower or flower cluster; as in *Allium*.

Segment – A division of a compound leaf or of a perianth.

Sepal – One of the members of the calyx.

Serrate – With teeth projecting forward.

Serrulate – Finely serrate.

Sessile – Lacking a petiole or stalk.

Silique – A dry elongated fruit divided by a partition between the two carpels.

Sinuate – With long wavy margins.

Sinus – A depression or notch in a margin between two lobes.

Sorus – The brown colored fruiting structure of ferns, often on the underside of the frond.

Spadix – Fleshy flower stalk bearing many tiny flowers as in arum family

Spathe – specialized bract enclosing the flower(s) as in arum family

Spatulate – Widened at the top like a spatula.

Spike – An elongated inflorescence with sessile (stalkless) or nearly sessile flowers.

Spikelet – A small or secondary spike; the ultimate flower cluster of the inflorescence of grasses and sedges.

Spine – A short thorn-like structure.

Spinose – With spines.

Spinulose – With small, sharp spines.

Stamen – Male reproductive structure of a flower, consisting of the pollen bearing structure (anther) borne on a stalk or filament.

Staminate – Male-flowered, with stamens only.

Standard – The large petal that stands up at the back of the flower as in a pea flower.

Stellate – Star-shaped.

Stipule – An appendage at the base of a leaf, or other plant part.

Stolon – A basal branch rooting at the nodes.

Striate – Marked with fine, longitudinal, parallel lines, ridges, or grooves.

T

Taproot – A strong, fleshy root that grows vertically into the soil, with smaller lateral roots.

Tendril – Thread-like stem or leaf that clings to adjacent structures for support (peas).

Tomentose – Densely matted with soft hairs.

Trifoliate – A compound leaf with three leaflets (clover).

Tuber – Swollen underground stem for storing food (potato, poison hemlock), that can sprout to form new plants. Tuberous - Forming tubers.

U

Umbel – Umbrella-shaped inflorescence, in which the pedicels (flower stalks) radiate from a common point like the ribs of an umbrella.

Undulate – Wavy, as the margins of leaves.

V

Veins – The vascular portions of the leaves.

Villous – Covered with short, fine hairs.

Viscid – Sticky.

W

Whorled – Three or more leaves, petals, or branches arranged in a ring at a node.

Wing – A thin, membranous extension of an organ.

Asclepiadaceae family, 38-40, 57-58, 92-93, 140
Asclepias poisoning, 38-40
Ascorbic acid, 5
Asparagus poisoning, 41
Asphyxiation, in Dieffenbachia poisoning, 105
Asteraceae family, 6-7, 117-118, 172, 250-252
Ataxia
 in Clivia poisoning, 78
 in Corydalis poisoning, 86-87
 in Crassula poisoning, 90
 in Crinum poisoning, 91
 in Dracaena poisoning, 108-109
 in Eucharis poisoning, 114
 in Galanthus poisoning, 124
 in Haemanthus poisoning, 129
 in Hippeastrum poisoning, 138
 in Hoya poisoning, 140
 in Laburnum angyroides poisoning, 164
 in Ligustrum poisoning, 173
 in Macadamia poisoning, 182
 in Melaleuca poisoning, 183
 in Melia azedarach poisoning, 185
 in Narcissus poisoning, 194
 in Oxalis poisoning, 204
 in Papaver poisoning, 206
 in Portulaca poisoning, 226
 in Sophora poisoning, 258-259
 in Sprekelia formossisima poisoning, 261
 in Taxus poisoning, 267
 in Zephyranthes poisoning, 291
Atropa belladonna poisoning, 42
Atropine
 for Bowiea poisoning, 48
 for Crassula poisoning, 90
 in Hyoscyamus poisoning, 148-149
 for Kalanchoe poisoning, 160
 for Lyonia poisoning, 180
 for oleander poisoning, 198
 for Pieris poisoning, 218
 for Rhododendron poisoning, 237
 in Solandra poisoning, 254-255
 for Theyetia poisoning, 273
Aucuba japonica poisoning, 43-44
Aucubacceae (Cornaceae) family, 43-44
Aucubin, 43
Australian umbrella tree, 246
Autumn crocus, 78-79
Avocado, 209-211
Azaleas, 12-13, 235-237

B
Balsam pear, 188
Bamboo, 191-192
Bane berries, 11-12
Bane berry, 11
Baptisia poisoning, 44-45
Barbados nut, 154-155
Barbaloin, 29-30
Begonia poisoning, 45-46
Begoniaceae family, 45-46
Behavioral abnormalities
 in Cannabis sativa poisoning, 62
 in Melaleuca poisoning, 183
 in Nepeta cataria poisoning, 195
 in Solandra poisoning, 254-255
Belarmine, 31-32
Belladonna, 42
Belladonna lily, 31-32
Benzofurans, 117-118
Benzophenanthridine alkaloids, 243
Berberidaceae family, 191-192, 224-225
Berberine
 in Chelidonium maius poisoning, 72-73
 in Nandina domestica poisoning, 191-192
Bersalgenins, 159-161
Beta-aminoproprionitile (BAPN), 168-169
Biliary hyperplasia
 in Heliotropium poisoning, 132
 in Lantana poisoning, 166-167
 in Ligularia poisoning, 172
 in Senecio poisoning, 252
 in Symphytum poisoning, 263-264
 in Zamia poisoning, 287
Biliary obstruction, 19
Biliary stasis, 166-167
Bird of paradise, 54
Birth defects, in Conium maculatum poisoning, 82-83
Bittersweet poisoning, 68-69
Black henbane, 148
Black locust, 240
Black locust poisoning, 240-241
Black walnut poisoning, 156-158
Bleeding hear, 102
Bleeding heart, 102-103
Blighia poisoning, 47
Blindness/partial blindness
 in Leucothoe poisoning, 171
 in Lyonia poisoning, 180
 in Ornithogallum poisoning, 202-203
 in Pieris poisoning, 218

Glory Lilly poisoning, 79
Glucosides
  in *Aesculus* poisoning, 15-16
  in *Buxus* poisoning, 53
Glucosinolates
  in *Armoracia (Cochlearea armoracea)* poisoning, 36-37
  names and toxic effects of, xxxiii
Glycoalkaloids
  in *Achillae* species, 6
  in *Cestrum* poisoning, 70-71
  in *Physalis* poisoning, 215
  plants containing, 255-256
  in *Solanum pseudocapsicum* poisoning, 255-256
Glycoproteins
  in *Phoradendron* poisoning, 213-214
  in *Ricinus communis* poisoning, 238-239
  in *Robinia* poisoning, 240-241
  in *Tulipa* poisoning, 277-278
  in *Wisteria* poisoning, 282
Glycosides
  in *Acokanthera*, 8
  in *Actaea*, 11-12
  in *Adenium*, 12-13
  in *Adonis*, 14
  in anemone, 32-33
  in *Caltha* poisoning, 59
  in *Cestrum* poisoning, 70-71
  in *Clematis* poisoning, 76
  in *Coriaria* poisoning, 85
  in *Corynocarpus* poisoning, 88
  in *Cycas* poisoning, 94
  in *Cyclamen* poisoning, 95-96
  in *Dracaena* poisoning, 108-109
  in *Eriobotrya* poisoning, 111-112
  in *Gymnocladus dioica* poisoning, 128
  in *Helleborus* poisoning, 133-134
  in *Hydrangea* poisoning, 145-146
  in *Ilex* poisoning, 150
  in *Kalanchoe* poisoning, 159-161
  in *Ligustrum* poisoning, 173
  names and toxic effects of, xxx-xxxi
  in *Nandina domestica* poisoning, 191-192
  in *Phytolacca* poisoning, 216
  in *Podophyllum* poisoning, 224-225
  in *Prunus* poisoning, 227-229
  in *Pyracantha* poisoning, 230
  in *Ranuculus* poisoning, 233
  in *Sambucus* poisoning, 242
  in *Scilla* poisoning, 249-250

  in *Taxus* poisoning, 266-268
  in *Thalictrum* poisoning, 269
  in *Theyetia* poisoning, 272-273
  in *Tulipa* poisoning, 277-278
  in *Zamia* poisoning, 286-287
Golden chain tree, 163-164
Golden shower, 67
Goose foot, 265
Grape vine, 280-281
Grayanotoxins
  in *Kalmia* poisoning, 162
  in *Leucothoe* poisoning, 170-171
  in *Lyonia* poisoning, 180
  in *Pieris* poisoning, 217-218
  plants containing, 162, 170, 180, 218, 236
  in *Rhododendron* poisoning, 235-237
Ground cherries, 215
Guernsey lily, 195-196
Gymnocladsapponins, 128
Gymnocladus dioica poisoning, 127-128

H
Habenero peppers, 64, 65
Haemanthamine
  in *Crinum* poisoning, 91
  in *Hippeastrum* poisoning, 137-138
Haemanthidine, 129
Haemanthus poisoning, 129
Hair loss, in *Mimosa* poisoning, 187
Hallucinations
  in *Atropa belladonna* poisoning, 42
  in *Brugmansia* poisoning, 49-50
  in *Coriaria* poisoning, 85
  in *Datura* poisoning, 99
  in *Ipomoea* poisoning, 151-152
  in *Nepeta cataria* poisoning, 195
  plants causing, xliii-xliv
  in *Solandra* poisoning, 254-255
  in *Sophora* poisoning, 259
Hashish poisoning, 61-62
Headache, in *Wisteria* poisoning, 282
Heart block
  Acokanthera in, 8
  in *Adenium obesum* poisoning, 13
  in *Bowiea* poisoning, 48
  in *Digitalis* poisoning, 107
  in *Kalanchoe* poisoning, 160
  in oleander poisoning, 197-198
  in *Theyetia* poisoning, 273
Heath family, 161-163, 179-181, 217-218, 235-237

Pulmonary edema, 157-158
Purgatives
in aloe, 29-30
Calotropois in, 57
for *Cannabis sativa* poisoning, 62
for *Humulus lupulus* poisoning, 142
Purkinje cell loss, 126
Purslane, 226
Pyracantha poisoning, 230
Pyridine alkaloids
in *Conium maculatum* poisoning, 82-83
in *Lobelia* poisoning, 176-177
in *Nicotiana* poisoning, 200-201
Pyridoxine, 21
Pyridoxine inhibitors
in *Albizia*, 20-21
in *Mimosa* poisoning, 186-187
Pyrroles, 172
Pyrrolizidine alkaloids
in *Heliotropium* poisoning, 131-132, 132
in *Ligularia* poisoning, 172
plants containing, 251
in *Senecio* poisoning, 251-252
in *Symphytum* poisoning, 263-264

Q

Quercus poisoning, 231-232
Quinolizidine alkaloids
in *Baptisia* poisoning, 44-45
in *Laburnum angyroides* poisoning, 163-164
plants containing, 164
in *Sophora* poisoning, 258-259

R

Rain lily, 290-291
Raisins, 280-281
Ranuculus poisoning, 232-233
Ranunculaceae family, 9-10, 11-12, 13-14,
32-33, 59, 76-77, 100-101, 133-134,
232-233, 269
Ranunculin
in *Actaea*, 11-12
in anemone, 32-33
in *Caltha* poisoning, 59
in *Clematis* poisoning, 76
in *Helleborus* poisoning, 133-134
in *Ranuculus* poisoning, 233
in *Thalictrum* poisoning, 269
Raphides. *See* Calcium oxalate crystals (raphides)
Recumbency
in *Lathyrus* poisoning, 169

in *Ligustrum* poisoning, 173
in *Oxalis* poisoning, 204
in *Portulaca* poisoning, 226
Red blood cell destruction, equine, 4
Red maple poisoning, 4-5
Renal failure
Acer toxin in, 5
in *Hemerocallis* poisoning, 135
in *Lilium* poisoning, 174-175
in *Oxalis* poisoning, 204
in *Philodendron* poisoning, 212
plants causing, xli
in *Portulaca* poisoning, 226
in *Quercus* poisoning, 232
in *Vitis* poisoning, 281
Renal tubular degeneration, in *Sesbania*
poisoning, 253
Renal tubular necrosis
in *Hemerocallis* poisoning, 135
in *Lilium* poisoning, 175
Reserpine, 279
Resins
in *Humulus lupulus* poisoning, 141-142
in *Sambucus* poisoning, 242
Respiratory failure
in *Asclepias* poisoning, 39-40
in *Atropa belladonna* poisoning, 42
in *Brugmansia* poisoning, 49-50
in *Gelsemium* poisoning, 125-126
Respiratory paralysis, 74-75
Respiratory signs. *See also* Breathing difficulty;
Respiratory failure
in *Armoracia (Cochlearea armoracea)*
poisoning, 36-37
in *Conium maculatum* poisoning, 82-83
in *Kalmia* poisoning, 162
in *Leucothoe* poisoning, 171
in *Lobelia* poisoning, 177
in *Lyonia* poisoning, 180
in *Papaver* poisoning, 206
in *Pieris* poisoning, 218
in *Rhododendron* poisoning, 236-237
Rhamnaceae family, 234
Rhamnus poisoning, 234
Rhododendron poisoning, 161-162, 180, 235-237
Rhodotoxin. *See also* Grayanotoxins
in *Kalmia* poisoning, 162
in *Leucothoe* poisoning, 170-171
Ricin, 155, 213
in *Ricinus communis* poisoning, 238-239
Ricinoleic acid, 238-239

T - #0364 - 101024 - C374 - 229/152/20 - PB - 9781591610281 - Gloss Lamination